General Chemistry I

2ND EDITION | LABORATORY MANUAL

CHEM 1141

Joel Gray, PhD and Cory Holland, PhD
Department of Chemistry and Biochemistry
Texas State University

General Chemistry I

CHEM 1141 Laboratory Manual
2nd Edition
Joel Gray, PhD and Cory Holland, PhD
Department of Chemistry and Biochemistry
Texas State University

Copyright © by the Department of Chemistry and Biochemistry
Copyright © by Van-Griner Learning

All products used herein are for identification purposes only, and may be trademarks or registered trademarks of their respective owners.

All rights reserved. No part of this book may be reproduced or transmitted in any form or by any means, electronic or mechanical, including photocopying, recording or by any information storage and retrieval system, without written permission from the author and publisher.

The information and material contained in this manual are provided "as is," and without warranty of any kind, expressed or implied, including without limitation any warranty concerning accuracy, adequacy, or completeness of such information. Neither the authors, the publisher nor any copyright holder shall be responsible for any claims, attributable errors, omissions, or other inaccuracies contained in this manual. Nor shall they be liable for damages of any type, including but not limited to, direct, indirect, special, incidental, or consequential damages arising out of or relating to the use of such material or information.

These experiments are designed to be used in college or university level laboratory courses, and should not be conducted unless there is an appropriate level of supervision, safety training, personal protective equipment and other safety facilities available for users. The publisher and authors believe that the lab experiments described in this manual, when conducted in conformity with the safety precautions described herein and according to appropriate laboratory safety procedures, are reasonably safe for students for whom this manual is directed. Nonetheless, many of the experiments are accompanied by some degree of risk, including human error, the failure or misuse of laboratory or electrical equipment, mis-measurement, spills of chemicals, and exposure to sharp objects, heat, blood, body fluids or other liquids. The publisher and authors disclaim any liability arising from such risks in connection with any of the experiments in the manual. Any users of this manual assume responsibility and risk associated with conducting any of the experiments set forth herein. If students have questions or problems with materials, procedures, or instructions on any experiment, they should always ask their instructor for immediate help before proceeding.

Printed in the United States of America
10 9 8 7 6 5 4 3 2 1
ISBN: 978-1-64565-257-1

Van-Griner Learning
Cincinnati, Ohio
www.van-griner.com

President: Dreis Van Landuyt
Project Manager: Janelle Krugh
Customer Care Lead: Lauren Wendel

Gray 65-257-1 F22
330084
Copyright © 2024

Table of Contents

Experiment #1	The Best Glassware for the Job: Using Density to Critically Evaluate Glassware	1
Experiment #2	That's Lit: Preparing and Evaluating Solutions	13
Experiment #3	All Mixed Up: Separating Components of Mixtures	25
Experiment #4	Copper Cycle: The Journey is Just as Important as the Destination	35
Experiment #5	How Much is there? A Simple Acid-Base Titration	47
Experiment #6	Titration^2: The Remix	55
Experiment #7	Metal Reactivities: Single & Double Displacement Reactions	67
Experiment #8	Acid-Base Neutralization: It's all about perspective	75
Experiment #9	Obey Hess's Law: It Can't be Broken	85
Experiment #10	This Experiment? It's a Gas!	93
Chemistry & Biochemistry Stockroom Supply Forms		101

CHEMISTRY 1141
GENERAL CHEMISTRY LABORATORY

Experiment #1
The Best Glassware for the Job:
Using Density to Critically Evaluate Glassware

Warnings and Hazards

The risks and hazards during this first experiment are minor however any time that you are in a lab, you must be cautious. The best way to remain safe is to understand RAMP.

Recognize hazards,
Assess risks,
Minimize risks, and
Prepare for emergencies

Using the space below, comment on each of the portions of the RAMP analysis below.

Recognize hazards (what are potential hazards that will be encountered in the procedure):
_____
_____
_____
_____

Assess risks (how can these hazards be harmful to people or your environment):
_____
_____
_____
_____
_____
_____

Minimize risks (what can you do to limit risk yourself and your environment):
_____
_____
_____
_____
_____

Prepare for emergencies. **Emergency: 911**
Stockroom: 512-245-3118

Background

Accuracy and precision are essential for any sort of scientific research. Working in a lab setting brings tons of new pieces of glassware. While it may appear that some many pieces of glassware are capable of collecting the same volumes - for instance, a 100mL beaker and 100mL graduated cylinder are all capable of holding 100mL of a solution, these have different levels of precision and accuracy.

Sometimes, you will to collect an exact volume of a solution but other times the volume that you collect may not matter. For example, if you are trying to cool a drink with an ice cube, you do not really care about the exact dimensions of that ice cube, you just want it to be cold. Meanwhile, if you are trying to make something like Sweet Tea, you will very carefully measure each ingredient. Each reactant in a reaction needs to be carefully measured but the overall reaction volume may not be nearly as important. Differing degrees of precision are needed in scientific experimentation and therefore different pieces of glassware are needed.

A beaker or an erlenmeyer flask are both examples of pieces of glassware that precision is in general somewhat less important. Meanwhile a piece of glassware like a volumetric flask is intended to be more precise. While these pieces of glassware are designed in a specific way, we will use today's experiment to determine the accuracy and precision of them. We will use the density of water as our standard - the density of water at room temperature is theoretically 1.00g/mL. The table on the next page shows all relevant information for this experiment.

GENERAL CHEMISTRY 1

	0.0	0.1	0.2	0.3	0.4	0.5	0.6	0.7	0.8	0.9
19	0.998405	0.998385	0.998365	0.998345	0.998325	0.998305	0.998285	0.998265	0.998244	0.998224
20	0.998203	0.998183	0.998162	0.998141	0.99812	0.998099	0.998078	0.998056	0.998035	0.998013
21	0.997992	0.99797	0.997948	0.997926	0.997904	0.997882	0.99786	0.997837	0.997815	0.997792
22	0.99777	0.997747	0.997724	0.997701	0.997678	0.997655	0.997632	0.997608	0.997585	0.997561
23	0.997538	0.997514	0.99749	0.997466	0.997442	0.997418	0.997394	0.997369	0.997345	0.99732
24	0.997296	0.997271	0.997246	0.997221	0.997196	0.997171	0.997146	0.99712	0.997095	0.997069

Table 1. The above table shows the density of water at temperatures ranging from 19 to 24 degrees celsius. From the <u>Handbook of Chemistry and Physics</u>, 53rd Edition.

> Example: The table above is complicated though has a huge amount of information. To determine the density of water at a specific temperature, say 22.7 degrees celsius, look at the far left column to find the whole number temperature and then look at the top row to find the decimal. To illustrate this, 22, .7, and 0.997608 are all greyed.

Working in Lab

Water: In almost every single lab that you work in, you'll find a pair of water faucets - one which dispenses tap water and one that dispenses deionized water. You will need to use the **tap** water but it will be limited to the initial rinsing of glassware when you are cleaning it. The **deionized** water on the other hand will be used for experimentation and final rinsing of glassware when cleaning. The difference between these two water sources is that there are dissolved minerals and ions in the tap water but not in the deionized water.

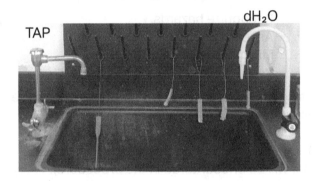

Digital balance: Depending on the lab that you're working in, your digital balance that you are using is very important, not so much for accuracy but instead precision. The two digital balances shown to the right display different numbers of significant figures and show different numbers of digits to the right of the decimal place.

The balances to the right show two (top) and three (bottom) numbers to the right of the decimal and therefore can conclusively and precisely declare a mass down to the thousandths position. The top balance only shows to the hundredths position but the bottom balance shows to the thousandths.

Significant Figures

Rule 1: All non-zero digits are ALWAYS significant (ex. 1, 2, <u>9</u>)
Rule 2: Any zero between two non-zero digits are significant (ex. 1<u>00</u>9)
Rule 3: Leading zeroes are NEVER significant (ex. 0.005)
Rule 4: Final and trailing zeroes are only significant if they follow a decimal (ex. 100 vs. 100.00)

Significant Figures

Addition & Subtraction: Round your final answer to the have the same number of decimal places as the number from your original numbers with the fewest number of decimal places.

Examples

34.325 - 1.26893 = 33.050607
Correctly rounded to 33.056

3.6 + 1.940257 = 5.540257
Correctly rounded to 5.5

22.5 + 13.694 = 36.194
Correctly rounded to _____

Multiplication & Division: Round your final answer to have the same number of significant figures as the number from your original calculation with the fewest number of significant figures.

6.29 / 1.6 = 3.93125
Correctly rounded to 3.9

13.9 * 32.395 = 450.2905
Correctly rounded to 450

127.44/10.6 = 12.0226415
Correctly rounded to _____

Scientific Notation

This practice should be your friend, using scientific notation is something that you should get comfortable with and use it to make writing numbers easier. Specifically, scientific notation is used for writing very large numbers and very small numbers. In addition, the guidelines for significant figures are applied when writing numbers using scientific notation. The general form for a number is shown below:

$$N \times 10^m$$

The number N is the signficant figures and m is the magnitude of the number or the number of digits the decimal place is moved.

The number N should be written ranging from 1 to 10 and all other significant digits should be written to the right of the decimal.

Positive exponent: Used when a number (N) is larger than 1.

Negative exponent: Used when a number (N) is less than 1.

Examples

0.325
Rewritten as 3.25×10^{-1}

0.0000004590
Rewritten as 4.590×10^{-7}

1000
Rewritten as 1×10^3

1000.0
Rewritten as 1.0000×10^3

602000000000000000000000
Rewritten as 6.02×10^{23}

1020.90
Correctly rewritten as _____

4.320×10^{-4}
Converted *from* scientific notation as

_____

Significant Figures Examples

Considering what you've just reviewed about significant figures as well as using significant figures in an equation, look at the pictures to the below.

Rewrite the number(s) in the display in scientific notation, if needed.

How many signficant figures are in the displays shown below?

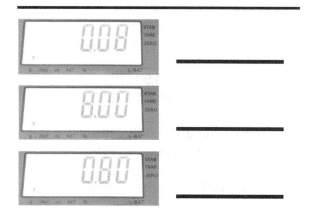
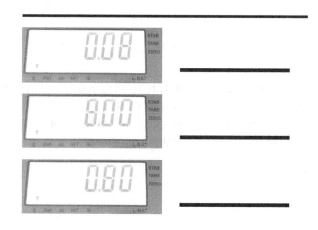

Other Measurements & Significant Figures

Taking measurements with a digital balance and determining the number of significant figures that the instrument has is fairly simple and direct. Determining the number of significant figures to use on other types of instruments or glassware is a little more complicated. However as a general rule, what lines/markings are on the glassware - you can reliably make a prediction of one additional number. To the right is a snapshot of a liquid that you might see on a graduated cylinder. It is easy to tell that the curved line (the meniscus) is higher than 10mL and less than 20mL. You can narrow that down more and conclude that the volume is greater than 15mL but less than 16mL - this is where you can make one small step and make an estimation of the volume to the tenths position. Making a prediction into the hundredths position would be making an estimation on top of another estimation which would be problematic.

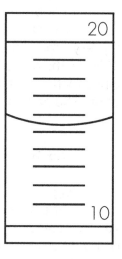

What is the approximate volume of the solution in the graduated cylinder shown above?

Significant Figure Practice Exercises
Calculate the density of a solution based on the volume displayed in the graduated cylinder and the mass displayed on the digital balance. The units in the graduated cylinder are mL and the digital balance units are grams. For your answer, be sure to use the correct number of significant figures, use the correct scientific notation, and record the correct units.

Significant Figures
When you are doing calculations with significant figures, keep in mind that the instruments that you are using guide your number of significant figures. For instance a digit balalnce like the one we're using here, you are only shown two numbers beyond the decimal place. If the instrument reads "13.65" that is because it is telling you its limitations - it would be inaccurate to interpret "13.65" as "13.650" or "13.651."

When the instrument or piece of glassware shows you a number, believe it, it's not lying to you!

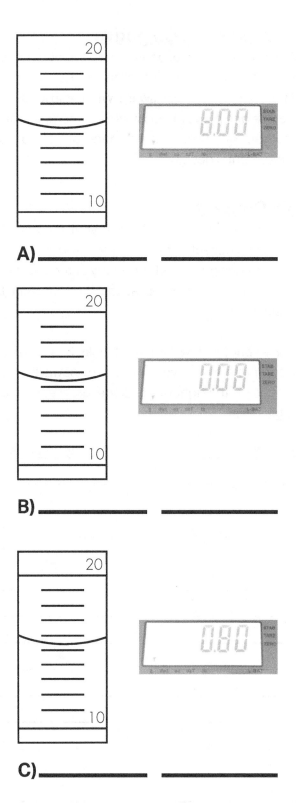

A) _____ _____

B) _____ _____

C) _____ _____

Experimental Objective

The point of this experiment is for you to gain a greater understanding about the precision of different pieces of glassware. When you know how precise one piece of glassware is in comparison to another, you will make better decisions about your experimental designs.

Bin Contents

All of your glassware will be in your bin and that bin will be collected from the stockroom after you have successfully filled out a copy of the glassware pickup document found at the back of your lab manual.

In the space below, fill in the blanks below to have a complete list of your week's glassware. In the spaces on the right of this page, draw each piece of glassware.

Glassware	Quantity	Capacity

Safety Notice

Gloves and goggles must be worn at all times when you are working in a lab. Take a moment before you begin working to identify where the eye wash station, the safety shower, and the exit are all located - in the box on the left, draw a simple model of the lab, marking where the shower and wash station are located (5pts).

Procedure
note, you will complete three identical trials for each piece of glassware

1. Fill your largest beaker with deionized water - the volume does not matter, you will be pouring *from* this beaker into other pieces of glassware. Determine the water temperature and record to the left.
2. Any time that you use the digital balance, use the tare function to zero out or reset the instrument.
3. Determine the mass of the empty 100mL beaker, record that mass in the data sheet on the next page, use the correct number of significant figures when recording the mass and all other numbers.
4. Carefully add 100mL of deionized water to the 100mL plastic beaker, use a plastic transfer pipette to transfer small volumes of deionized water to get as close as possible to the 100mL level. The bottom of the meniscus should be on the 100mL mark on the beaker.
5. As soon as you have precisely transferred 100mL of deionized water to the beaker, using the same digital balance that you previously used, determine the mass of the beaker with 100mL of deionized water.
6. At this point you have recorded the two experimental values that you need to collect for trial 1. Empty and dry your 100mL beaker, repeat steps 3-5 two more times and fill in the spaces for mass of empty and full beaker for trials 2 & 3.
7. Proceed to the next page to complete the calculations for each trial.

Water temperature _____

Density of water _____
see table on page 2

Use the space below to draw a flow chart of the steps to the right

Calculations

The calculations will be the same for all pieces of glassware. Complete calculations for each trial individually, so finish trial 1 then move on to trial 2, and so on.

1. To calculate the mass of water, simply subtract the mass of the empty beaker from the mass of the full beaker.

2. The theoretical volume of H_2O is what you attempted to add to the beaker. In this case you added water up to the 100mL line, which means you attempted to add 100mL of deionized water to the beaker, therefore your theoretical volume is 100mL.

3. The experimental volume of H_2O is a little more complicated. The experimental volume of H_2O is deteremined using the mass of water and the density of the water. You recorded the density of water based on the temperature from the table on page 2. Units for density are (g/mL) and units for mass are (g) - to determine the experimental volume, use the mass of water and divide it by the density.

4. The last calculation for each trial is the percent error calculation. A percent error calculation is generally recorded as a positive value which is why the lines for absolute value are used below. This calculation compares your experimental result to the theoretical value.

$$\frac{|(\text{theoretical})-(\text{experimental})|}{(\text{theoretical})} \times 100 = \underline{\qquad}$$

5. The final calculation gives you the average of your three percent error values. This value can be viewed as the accuracy of this piece of glassware any time that you use it.

Trial 1
Mass of empty beaker _____
Mass of full beaker _____
Mass of water _____
Theoretical vol H_2O _____
Experimental vol H_2O _____
Percent Error _____

Trial 2
Mass of empty beaker _____
Mass of full beaker _____
Mass of water _____
Theoretical vol H_2O _____
Experimental vol H_2O _____
Percent Error _____

Trial 3
Mass of empty beaker _____
Mass of full beaker _____
Mass of water _____
Theoretical vol H_2O _____
Experimental vol H_2O _____
Percent Error _____

Average Percent Error
T1 % Error _____
T2 % Error _____
T3 % Error _____
Average Percent Error _____

Procedure

8. Refill your largest beaker with deionized water - the volume does not matter, you will be pouring *from* this beaker into other pieces of glassware.
9. Any time that you use the digital balance, use the tare function to zero out or reset the instrument.
10. Determine the mass of the empty 125mL erlenmeyer flask (also known as "e. flask"), record that mass in the data sheet on the next page, use the correct number of significant figures when recording the mass and all other numbers.
11. Carefully add 100mL of deionized water to the 125mL e. flask, use a plastic transfer pipette to transfer small volumes of deionized water to get as close as possible to the 100mL level. The bottom of the meniscus should be on the 100mL mark on the beaker.
12. As soon as you have precisely transferred 100mL of deionized water to the e. flask, using the same digital balance that you previously used, determine the mass of the e. flask with 100mL of deionized water.
13. At this point you have recorded the two experimental values that you need to collect for trial 1. Empty and dry your 125mL e. flask, repeat steps 10-12 two more times and fill in the spaces for mass of empty and full e. flask for trials 2 & 3.
14. Complete your calculations as you previously did.
15. The two remaining pieces of glassware - the graduated cylinder (also known as "grad cyl") and the volumetric flask (also known as "v. flask") will be evaluated the same way as the beaker and e. flask.
16. Repeat steps 10-14, with the same volumes, and fill the data in on the next page.

Trial 1
Mass of empty e. flask _____
Mass of full e. flask _____
Mass of water _____
Theoretical vol H_2O _____
Experimental vol H_2O _____
Percent Error _____

Trial 2
Mass of empty e. flask _____
Mass of full e. flask _____
Mass of water _____
Theoretical vol H_2O _____
Experimental vol H_2O _____
Percent Error _____

Trial 3
Mass of empty e. flask _____
Mass of full e. flask _____
Mass of water _____
Theoretical vol H_2O _____
Experimental vol H_2O _____
Percent Error _____

Average Percent Error
T1 % Error _____
T2 % Error _____
T3 % Error _____
Average Percent Error _____

Trial 1		Trial 1	
Mass of empty grad cyl	_____	Mass of empty v. flask	_____
Mass of full grad cyl	_____	Mass of full v. flask	_____
Mass of water	_____	Mass of water	_____
Theoretical vol H_2O	_____	Theoretical vol H_2O	_____
Experimental vol H_2O	_____	Experimental vol H_2O	_____
Percent Error	_____	Percent Error	_____

Trial 2		Trial 2	
Mass of empty grad cyl	_____	Mass of empty v. flask	_____
Mass of full grad cyl	_____	Mass of full v. flask	_____
Mass of water	_____	Mass of water	_____
Theoretical vol H_2O	_____	Theoretical vol H_2O	_____
Experimental vol H_2O	_____	Experimental vol H_2O	_____
Percent Error	_____	Percent Error	_____

Trial 3		Trial 3	
Mass of empty grad cyl	_____	Mass of empty v. flask	_____
Mass of full grad cyl	_____	Mass of full v. flask	_____
Mass of water	_____	Mass of water	_____
Theoretical vol H_2O	_____	Theoretical vol H_2O	_____
Experimental vol H_2O	_____	Experimental vol H_2O	_____
Percent Error	_____	Percent Error	_____

Average Percent Error		Average Percent Error	
T1 % Error	_____	T1 % Error	_____
T2 % Error	_____	T2 % Error	_____
T3 % Error	_____	T3 % Error	_____
Average Percent Error	_____	Average Percent Error	_____

Reflections

Answer the questions below using your data - critically evaluate and analyze your results when answering these questions.

1. Which piece of glassware had the lowest average percent error and which had the highest average percent error?

2. Were your results what you expected or did the average percent errors of the different pieces of glassware surprise you?

3. What chacteristic(s) about the shape of the most accurate piece of glassware possibly contributed to the accuracy? (Is there something you noticed about the shape of the most accurate glassware that makes it distinct from the less accurate pieces?)

4. Why was it necessary to do three trials for each piece of glassware?

5. Do you think that your accuracy would have changed at all if ethanol (density ~ 0.79g/mL) was used instead of deionized water?

Safety Data Sheets

The QR code shown below will take you to the Safety Data Sheet for the chemical(s) used in this experiment

Water

CHEMISTRY 1141
GENERAL CHEMISTRY LABORATORY

Experiment #2
That's Lit:
Preparing and Evaluating Solutions

Warnings and Hazards

The risks and hazards during this second experiment are minor however any time that you are in a lab, you must be cautious. The best way to remain safe is to understand RAMP.

Recognize hazards,
Assess risks,
Minimize risks, and
Prepare for emergencies

Using the space below, comment on each of the portions of the RAMP analysis below.

Recognize hazards (what are potential hazards that will be encountered in the procedure):
_____
_____
_____
_____

Assess risks (how can these hazards be harmful to people or your environment):
_____
_____
_____
_____

Minimize risks (what can you do to limit risk yourself and your environment):
_____
_____
_____
_____

Prepare for emergencies.
Emergency: 911
Stockroom: 512-245-3118

Background

Making a solution is one of the first things you need to understand how to do in a chemistry lab or any lab for that matter. Preparing a solution is a critical process are requires careful measurements and mixing. In most instances, the pH of a solution will need to be adjusted. In some cases solids will have differently based on the type of compound they are - which is what we will explore today.

Today we'll look at molecular and ionic compounds and how they go into solution. What we'll be looking at first and foremost is _does_ a substance go into solution? All of the substances that we look at we'll be asking first, do they dissolve in water? Whether or not a substance dissolves can simply be answered by the naked eye. A solution that appears to have a uniform appearance throughout indicates a homogeneous solution! If a substance dissolves, a word like soluble is used to describe that solid.

A solid which does not dissolve is described as insoluble - now what makes this a little bit complicated is that if a substance dissolves or does not dissolve, you do not get the complete story. In the case of ionic compounds, we are interested in solubility but we are also interested in an ionic compound's ability to dissociate. When an Ionic compound dissociates, the ions that make up the compound will separate from one another and exist as free ions in the solution. For example, if we take the solid, sodium chloride (NaCl) and added it to water we would observe this solid dissolve. We would see the solid more or less disappear as it dissolved into the water. Remember though, the important part of the story is what happens to those solids at the atomic and molecular levels. The NaCl will dissociate into individual components. The NaCl which has an overall charge of zero will separate into two charged species, also known as ions, Na^+ and Cl^-.

We will begin to look at substances as soluble or insoluble - we will also look at whether or not they dissociate. Now, dissociation is a bit complicated because that is on a molecular and atomic level. Thankfully we can evaluate a substance and determine if it dissociates by looking to see if it conducts electricity or not. If a substance conducts electricity then it can be said that it is an **electroltye**. If a substance does <u>not</u> conduct electricity, it can be said that it is a **non-electrolyte**. Every substance can be classified using this terminology. Think about this terminology as the two ends of a spectrum - meaning that this is not a question that has a binary answer. Not all substances are electrolytes or non-electrolytes, many substances are stronger electrolytes than non-electrolytes but are not strong electrolytes.

Ultimately it is common to see the terminology and descriptions used below.

1. Strong electrolyte will completely dissociate when dissolved in solution. **Strong electrolytes** will conduct electricity very well within a solution.

2. Weak electrolyte will partially dissociate when dissolved in solution. **Weak electrolytes** will weakly conduct electricity within a solution.

3. Nonelectrolytes will not dissociate when dissolved in solution. **Nonelectrolytes** will not conduct electricity at all within a solution.

Based on the description to the left, label each of the images as electrolytes, non-electrolytes, and weak electrolytes.

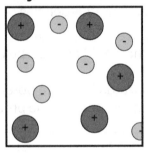

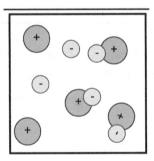

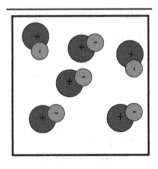

_____

Experimental Objective

The point of this experiment is for you to develop the technical skill of preparing a simple solution. This experiment will also illustrate the differences in solutions which dissolve, dissociate, and do neither!

Bin Contents

All of your glassware will be in your bin and that bin will be collected from the stockroom after you have successfully filled out a copy of the glassware pickup document found at the back of your lab manual.

In the space below, fill in the blanks below to have a complete list of your week's glassware. In the space on the right of this page, draw each piece of glassware.

Glassware	Quantity	Capacity
_____	_____	
_____	_____	
_____	_____	
_____	_____	
_____	_____	
_____	_____	
_____	_____	
_____	_____	
_____	_____	
_____	_____	
_____	_____	
_____	_____	
_____	_____	

Controls

"You never had control, that's the illusion!"
—Dr. Ellie Sattler

Unlike in the 1993 film, *Jurassic Park*, we will be able to control the system that we're working with. However we'll be using a different type of control - rather than exerting a power type of control, we'll be utilizing experimental controls. We will be including several different types of controls in our experiments. In addition to several different types we will try several different examples of those types of controls. Each control supports our overall experiment but in different ways - though with the overall idea that they will validate our experimental results.

Positive controls: A positive control is a test treatment that produces a known result. We have two positive controls for this experiment - the first of which will confirm that our battery and light system work. The second positive control will be a follow-up from our first positive control, this second positive control will incorporate a solution of HCl. This second positive control using HCl will validate not only that our battery/light system works but our battery/light system works with a strong electrolyte.

Negative controls: A negative control is a test treatment that produces a known result a well however in this case, the test should produce no result. We have two negative controls for this experiment - the first of which will simply confirm that deionized water does not conduct electricity - something needs to be dissolved in that solution. The second negative control is a solution of ethanol which will confirm that a molecular substance will not conduct electricity.

In the space below, draw a simple model of the positive controls and the negative controls, include the beaker with the solution, battery, and light system.

Positive control 1

Positive control 2

Negative control 1

Negative control 2

Safety Notice
Gloves and goggles must be worn at all times when you are working in a lab. Take a moment before you begin working to identify where the eye wash station, the safety shower, and the exit are all located - in the box on the left, draw a simple model of the lab, marking where the shower and wash station are located.

Procedure

Positive Control[1] – this control shows you what the experiment looks like if it works as expected

1. Collect your 9V battery, length of green wire, and your light system.
2. Connect one end of the green wire to the battery terminal labeled "+."
3. Connect the other end of that same green wire to one end of your light system.
4. Connect the free end of your light system to the battery terminal labeled "-."
5. Your light/lights should turn on, if they do not - consult your IA to ensure that you have connected the apparatus properly.

Positive Control[2] – this control shows you what the experiment looks like if it works as expected when using a solution

6. After confirming that your first positive control was successful, proceed to transfer a small volume of HCl and place it into a well in your well plate.
7. Carefully place one end of your free green wire in the solution that you just placed in the well plate.
8. Attach the other end of your green wire to the "+" terminal of your 9V battery.

Did your positive control[1] work? _____

Notes: _____

Did your positive control[2] work? _____

Notes: _____

GENERAL CHEMISTRY 1 | 17

9. Using your light system, connect one end to the "-" terminal of your 9V battery and place the other end in the solution with your other wire. *Do not let the wires come in contact in the solution.
10. Once the expected result has been confirmed, carefully place your wire and battery on a paper towel.

Negative Control[1] – this control shows you what the experiment looks like if no result is observed

1. After carefully cleaning out your well plate, add deionized water to one cell.
2. Connect one end of the green wire to the battery terminal labeled "+" and place the other end in the deionized water in the well plate.
3. Connect one end of the light system to the battery terminal labeled "-" and place the other end in the deionized water place.
4. Your light should not turn on - proceed if this result is observed, speak with your IA if it is not observed.
5. Discard the solution from your well plate and clean it out.

Did your negative control[1] work?

_____

Notes: _____
_____
_____

Negative Control[2] – this control shows you what the experiment looks like if no result is observed

1. Add ethanol to one well within your well place
2. Connect one end of the green wire to the battery terminal labeled "+" and the other into the ethanol solution.
3. Connect one end of your light system to the "-" terminal and the other in the ethanol.
4. Your light/lights should not turn on, if they do - consult your IA to ensure that you have connected the apparatus properly.

Did your negative control[2] work?

_____

Notes: _____
_____
_____

18 | GENERAL CHEMISTRY 1

Part I Data Set

Full name of NaCl   _____

Mass of NaCl (g)   _____

Molar mass of NaCl   **58.44g/mol**

# of Moles of NaCl   _____

Volume of solution (mL)   _____

Volume of solution (L)   _____

Molarity of NaCl solution (M) _____

Is NaCl soluble in dH_2O? _____

Dilution #1 Concentration (M) _____

Does D1 conduct electricity? _____

Dilution #2 Concentration (M) _____

Does D2 conduct electricity? _____

Does NaCl dissociate?   _____

Notes & Observations _____
_____
_____
_____
_____

Procedure

Part I:
<u>Working with NaCl: is it a strong electrolyte, weak electrolyte, or non-electrolyte?</u>

1. Collect ____g of NaCl - record the exact mass in the space provided to the left.
2. Transfer all of the NaCl to your 100mL volumetric flask.
3. Collect approximately 200mL of deionized water in your 250mL plastic beaker.
4. Transfer approximately 50mL of deionized water into the volumetric flask with the NaCl.
5. Gently swirl the volumetric flask until the NaCl goes into solution - should not take more than 30 seconds.
6. When the NaCl goes into solution, add more deionized water water to the 100mL marking on the volumetric flask.
7. After preparing this solution, use a plastic transfer pipette to transfer approximately 3-5mL of this solution to one well within your well plate.

   *Do not discard the remaining solution!
8. Now for your experimental trial - place one end of your green wire into the solution in your well plate, place the other end of your green wire into onto the "+" electrode on your battery.
9. Using your light array, place one end of the wire going to the light in the solution on your well plate and one on the "-" electrode on your battery.
10. Record your observations and fill in the blank spaces in the left on your data spaces - discard the well plate contents.

NaCl Dilutions
11. Carefully transfer 50mL from your volumetric flask into your 50mL graduated cylinder.
12. Discard the remaining solution in your volumetric flask and thoroughly rinse it.
13. Transfer all remaining 50mL from your graduated cylinder back into the volumetric flask.

14. Add more deionized water to the volumetric flask to adjust the total volume to 100mL - this solution is your first dilution, aka D1 - dilution #1.
15. Transfer approximately 3-5mL of D1 into the well plate,
16. Repeat steps 9 & 10 for this new solution and complete the data sheet on the previous page for D1 - discard the well plate contents.
17. Repeat steps 11-16 as described above, this will prepare your second dilution D2, record your observations and commentary.

Part II:
Working with $CaCO_3$: is it a strong electrolyte, weak electrolyte, or non-electrolyte?

1. Collect ____ g of $CaCO_3$ - record the exact mass in the space provided to the left.
2. Transfer all of the $CaCO_3$ to your 100mL volumetric flask.
3. Collect approximately 200mL of deionized water in your 250mL plastic beaker.
4. Transfer approximately 50mL of deionized water into the volumetric flask with the $CaCO_3$.
5. Gently swirl the volumetric flask until the $CaCO_3$ goes into solution - swirl for no more than 30 seconds. If the $CaCO_3$ does not go into solution after 30 seconds, draw a conclusion if this is soluble or insoluble.
6. Add the additional deionized water to adjust the volume in your volumetric flask to 100mL.
7. After preparing this solution, use a plastic transfer pipette to transfer approximately 3-5mL of this solution to one well within your well plate.

*Do not discard the remaining solution!

8. Now for your experimental trial - place one end of your green wire into the solution in your well plate, place the other end of your green wire into onto the "+" electrode on your battery.
9. Using your light array, place one end of the wire going to the light in the solution on your well plate and one on the "-" electrode on your battery.

Part II Data Set

Full name of $CaCO_3$ _____

Mass of $CaCO_3$ (g) _____

Molar mass of $CaCO_3$ 100.09g/mol

# of Moles of $CaCO_3$ _____

Volume of solution (mL) _____

Volume of solution (L) _____

Molarity of $CaCO_3$ solution (M) _____

Is $CaCO_3$ soluble in dH_2O? _____

Dilution #1 Concentration (M) _____

Does D1 conduct electricity? _____

Dilution #2 Concentration (M) _____

Does D2 conduct electricity? _____

Does CaCO3 dissociate? _____

Notes & Observations _____
_____
_____
_____
_____

Part III Data Set

Full name of $C_{12}H_{22}O_{11}$ _____

Mass of $C_{12}H_{22}O_{11}$ (g) _____

Molar mass of $C_{12}H_{22}O_{11}$ 342.30g/mol

# of Moles of $C_{12}H_{22}O_{11}$ _____

Volume of solution (mL) _____

Volume of solution (L) _____

Molarity of $C_{12}H_{22}O_{11}$ solution (M) _____

Is $C_{12}H_{22}O_{11}$ soluble in dH_2O? _____

Dilution #1 Concentration (M) _____

Does D1 conduct electricity? _____

Dilution #2 Concentration (M) _____

Does D2 conduct electricity? _____

Does $C_{12}H_{22}O_{11}$ dissociate? _____

Notes & Observations _____
_____
_____
_____
_____

10. Record your observations and fill in the blank spaces in the left on your data spaces - discard the well plate contents.

$CaCO_3$ Dilutions

11. Carefully transfer 50mL from your volumetric flask into your 50mL graduated cylinder.
12. Discard the remaining solution in your volumetric flask and thoroughly rinse it.
13. Transfer all remaining 50mL from your graduated cylinder back into the volumetric flask.

Part III:
Working with $C_{12}H_{22}O_{11}$: is it a strong electrolyte, weak electrolyte, or non-electrolyte?

1. Collect ____g of $C_{12}H_{22}O_{11}$ - record the exact mass in the space provided to the left.
2. Transfer all of the $C_{12}H_{22}O_{11}$ to your 100mL volumetric flask.
3. Collect approximately 200mL of deionized water in your 250mL plastic beaker.
4. Transfer approximately 50mL of deionized water into the volumetric flask with the $C_{12}H_{22}O_{11}$.
5. Gently swirl the volumetric flask until the $C_{12}H_{22}O_{11}$ goes into solution - swirl for no more than 30 seconds. If the $C_{12}H_{22}O_{11}$ does not go into solution after 30 seconds, draw a conclusion if this is soluble or insoluble.
6. Add the additional deionized water to adjust the volume in your volumetric flask to 100mL.
7. After preparing this solution, use a plastic transfer pipette to transfer approximately 3-5mL of this solution to one well within your well plate.

   *Do not discard the remaining solution!
8. Now for your experimental trial - place one end of your green wire into the solution in your well plate, place the other end of your green wire into onto the "+" electrode on your battery.

Reflections

Answer the questions below using your data - critically evaluate and analyze your results when answering these questions.

1. Classify each of your compounds (NaCl, $CaCO_3$, and $C_{12}H_{22}O_{11}$) as ionic or molecular.

2. Which of the compounds were soluble in dH_2O?

3. Which of the compounds in this experiment dissociated in dH_2O?

4. Was there a correlation between concentration and ability of the solution to conduct electricity?

5. How would you classify your compounds as strong electrolytes, weak, and non-electrolytes?

Safety Data Sheets

The QR code shown below will take you to the Safety Data Sheet for the chemical(s) used in this experiment

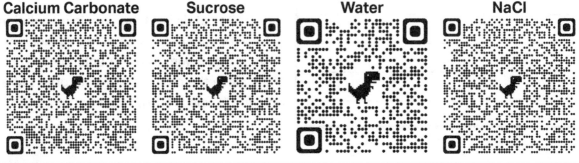

CHEMISTRY 1141
GENERAL CHEMISTRY LABORATORY

Experiment #3
All Mixed Up:
Separating Components of Mixtures

Warnings and Hazards

The risks and hazards during this second experiment are minor however any time that you are in a lab, you must be cautious. The best way to remain safe is to understand RAMP.

Recognize hazards,
Assess risks,
Minimize risks, and
Prepare for emergencies

Using the space below, comment on each of the portions of the RAMP analysis below.

Recognize hazards (what are potential hazards that will be encountered in the procedure):
_____
_____
_____
_____

Assess risks (how can these hazards be harmful to people or your environment):
_____
_____
_____
_____
_____

Minimize risks (what can you do to limit risk yourself and your environment):
_____
_____
_____
_____
_____

Prepare for emergencies.
  Emergency: 911
  Stockroom: 512-245-3118

Background

Separating the components of a mixture is an important technical skill to develop. When you have a mixture of solids, sometimes these substances can be separated using physical properties of the components and other instances you'll use a chemical property.

In this experiment we will explore physical changes as well as chemical changes to substances within different mixtures. We will also explore a separation technique common to most people on a daily basis - filtration. When you brew coffee, the ground beans are heated up and your coffee is mixed in with the beans. Obviously drinking the ground coffee beans would make for an uneasy experience so we filter the ground beans from the brewed coffee. Filtration can and is much more complex than a coffee filter but a coffee filter effectively does the trick.

When you use a filter, it will become saturated with a solution and as a result we do need to dry that filter paper. In this experiment we will explore three different drying strategies to determine which is most effective.

Experimental Objective
The point of this experiment is understand how to separate and isolate individual components of a mixture and to compare and evaluate the best way to quickly and efficiently dry a damp filter paper.

Bin Contents
All of your glassware will be in your bin and that bin will be collected from the stockroom after you have successfully filled out a copy of the glassware pickup document found at the back of your lab manual.

In the space below, fill in the blanks below to have a complete list of your week's glassware. In the space on the right of this page, draw each piece of glassware.

Glassware	Quantity	Capacity
_____	_____	
_____	_____	
_____	_____	
_____	_____	
_____	_____	
_____	_____	
_____	_____	
_____	_____	
_____	_____	
_____	_____	
_____	_____	
_____	_____	

Procedure
Part I:
Working with your coffee filter

Note: this portion of your experiment involves long periods of drying, during those waiting times, proceed to remainder of the experiment

1. Collect three identical coffee filters from the fume hood.
2. Carefully in half two times and then tear away the "baffles" (the ruffled, curvy portion of the coffee filter) and discard them.
3. With a sharpie, label these de-baffled coffee filters as "C1," "C2," and "C3."
4. Determine the dry mass of the de-baffled coffee filters and record that in the space to the left.
5. Collect the three watch glasses and label them with "C1," "C2," and "C3."
6. Determine the mass of the three watch glasses and record those masses in the space to the left (and on the next page).
7. Collect your erlenmeyer flask and funnel - place your funnel in the opening of the erlenmeyer flask.
8. Place "C1" filter paper in the funnel.
9. Collect ~100mL of deionized water and pour it through the "C1" filter paper, ensuring that the filter paper is completely saturated with water.
10. Place the "C1" filter paper on your pre-weighed "C1" watch glass.
11. Determine the mass of your saturated "C1" filter paper and the "C1" watch glass. Record this mass in the space to the left.
12. Place the damp watch filter paper + watch glass in the oven in your lab and let it incubate for 15, 30, and 45 minutes. At each time interval, determine the mass and record it in the space to the left.
13. Collect your erlenmeyer flask and funnel - place your funnel in the opening of the erlenmeyer flask.
14. Place "C2" filter paper in the funnel.
15. Collect ~100mL of deionized water and pour it through the "C2" filter paper, ensuring that the filter paper is completely saturated with water.

"C1" Data - Oven Drying
Mass of coffee filter "C1" _____g
Mass of watch glass "C1" _____g

Mass of watch glass, filter, & water _____g
Mass of water _____g

Mass of watch glass, filter, & water
  *after 15 min _____g
Mass of watch glass, filter, & water
  *after 30 min _____g
Mass of watch glass, filter, & water
  *after 45 min _____g

Mass of water still in filter paper after 45min
_____g

"C2" Data - EtOH + Oven
Mass of coffee filter "C2" _____g
Mass of watch glass "C2" _____g

Mass of watch glass, filter, & water _____g
Mass of water _____g

Mass of watch glass, filter, & water
  *after 15 min _____g
Mass of watch glass, filter, & water
  *after 30 min _____g
Mass of watch glass, filter, & water
  *after 45 min _____g

Mass of water still in filter paper after 45min
_____g

16. With the filter paper still in the funnel spray the filter paper with ethanol.

17. Transfer the ethanol sprayed filter paper to the "C2" watch glass.

18. Determine the mass of the ethanol sprayed "C2" filter paper and the "C2" watch glass and record that mass on the previous page.

19. Place the damp watch filter paper + watch glass in the oven in your lab and let it incubate for 15, 30, and 45 minutes. At each time interval, determine the mass and record it on the previous page.

20. Collect your erlenmeyer flask and funnel - place your funnel in the opening of the erlenmeyer flask.

21. Place "C3" filter paper in the funnel.

22. Collect ~100mL of deionized water and pour it through the "C3" filter paper, ensuring that the filter paper is completely saturated with water.

23. Transfer the filter paper to the "C3" watch glass.

24. Determine the mass of the "C3" filter paper and the "C3" watch glass and record that mass in the space to the right.

26. Place the damp watch filter paper + watch glass in a safe space near your working area where it will not be disrupted.

"C3" Data - Room Temperature
Mass of coffee filter "C2" _____ g
Mass of watch glass "C2" _____ g

Mass of watch glass, filter, & water _____ g
Mass of water _____ g

Mass of watch glass, filter, & water
 *after 15 min _____ g
Mass of watch glass, filter, & water
 *after 30 min _____ g
Mass of watch glass, filter, & water
 *after 45 min _____ g

Mass of water still in filter paper after
 45min _____ g

Part II:

<u>Separating iron from a quaternary mixture</u>

1. Collect a plastic beaker, determine the mass of the beaker and record it to the right for "beaker #1."

2. Collect approximately 2-3 g of the quaternary mixture, in your plastic beaker and record the precise mass in the space to the right.

3. Collect the other plastic beaker, determine the mass and record it in the space to the right for "beaker #2."

4. Place your magnet at the base of "beaker #1" and carefully pour the mixture from "beaker #1" to "beaker #2."

5. Determine the mass of "beaker #2" and the contents transferred from "beaker #1."

6. Determine the mass of iron from the original mixture.

Quaternary Mixture Data
Mass of Beaker #1 _____ g
Mass of Beaker + mixture _____ g
Mass of mixture _____ g
Mass of Beaker #2 _____ g

Mass of Beaker #2 + mixture _____ g

Mass of Iron _____ g

Quaternary Mixture Data
Mass of Erlenmeyer flask	_____ g
Mass of E flask + NaCl	_____ g
Mass of NaCl	_____ g

Part III:
Separating NaCl from a quaternary mixture

1. Collect a coffee filter and remove the baffles from the flask.
2. Collect approximately 50mL of deionized water in your graduated cylinder.
3. Transfer approximately 25mL deionized water into "beaker #2."
4. Determine the mass of your dry erlenmeyer flask and record it in the space to the left.
5. Place the funnel in the erlenemeyer flask and place the filter paper in the funnel.
6. Transfer the contents of "beaker #2" to the funnel - do not overload the filter paper and do not let any solution seep between the filter paper and the funnel.
7. Transfer additional deionized water to "beaker #2" as needed to transfer all of the materials from "beaker #2" to the funnel.
8. Carfully remove the funnel from the erlenmeyer flask and place the erlenmeyer flask on the hot plate and begin heating the contents of the erlenemeyer flask on the hot plate until all water has evaporated.
9. After the solution in the evaporating dish has evaporated, turn the hot plate off, allow the erlenmeyer flask to cool - determine the mass of the erlenmeyer flask + NaCl.

Part IV:
Separating sand from a quaternary mixture

1. Determine the mass of your evaporating dish and record in the space to the left.
2. Carefully transfer the filter paper from your funnel onto your evaporating dish.
3. Add HCl to the filter paper in your evaporating dish.
4. Add HCl to this mixture until no more bubbles develop.
5. After bubbles stop forming, carefully remove the filter paper from the evaporating dish - leaving behind the solution and any solids - discard the filter paper.

Quaternary Mixture Data
Mass of Evaporating dish	_____ g
Mass of E dish + sand	_____ g
Mass of sand	_____ g

6. Determine the mass of your glass beaker - record the mass in the space to the right.

7. Carefully decant the supernatant solution from your evaporating dish into the pre-weighed glass beaker.

8. Place your evaporating dish on the hot plate and allow it to heat until the solution has evaporated.

9. After the solution has evaporated, remove the evaporating dish from the hot plate and place it on a hot pad to cool to room temperature.

10. Determine mass of evaporating dish + sand and record in the space to the right.

Quaternary Mixture Data
Mass of Evaporating dish _____g
Mass of E dish + sand _____g

Mass of sand _____g

Part V:
Recovering $CaCO_3$ from a quaternary mixture

1. Place the beaker containing the supernatant solution on the hot plate - heat to a boil and carefully add the potassium carbonate solution.

2. Heat the supernatant solution + potassium carbonate for an additional 5 minutes, reduce heat as soon as it begins boiling.

3. Remove beaker from hot plate to cool.

4. Collect a coffee filter, remove and discard the baffles.

5. Determine the mass of the de-baffled coffee filter, determine the mass of watch glass and label with your initials.

6. Place your de-baffled filter in the funnel in the erlenmeyer flask.

7. Carefully pour contents of beaker into funnel + filter to collect white solid.

8. Transfer your filter paper to your watch glass.

9. Place watch glass in oven and incubate for at least 10 minutes - remove, determine the mass of the watch glass + filter paper + solid.

Quaternary Mixture Data
Mass of watch glass _____g
Mass of filter paper _____g

Mass of watch glass + filter paper + solid after 10min

_____g

Mass of filter paper + solid after 10 min

_____g

Mass of solid after 10min

_____g

Mass of $CaCO_3$ _____g

Notes: _____
_____
_____

Reflections

Answer the questions below using your data - critically evaluate and analyze your results when answering these questions.

1. In part I of your experiment, you tested three different methods for drying a dampened coffee filter - which method was best? Room temperature drying, drying in the oven, drying in the oven with ethanol?

2. From part I - is there anything that you would change or believe would have helped speed up the drying process?

3. What were the four components of your quaternary mixture?

4. Based on your results from the quaternary mixture separation, what is the percent composition of each component?

5. From your results of part II-V, do you think that there was any water residue in your samples?

6. In part III of the procedure, deionized water was used to separate the NaCl from the remaining mixture - had HCl been used in that step instead of deionized water, how would your results differ?

Safety Data Sheets
The QR code shown below will take you to the Safety Data Sheet for the chemical(s) used in this experiment

Water

Silicon Dioxide

Calcium Carbonate

HCl

Ethanol

Iron Filings

CHEMISTRY 1141
GENERAL CHEMISTRY LABORATORY

Experiment #4
Copper Cycle:
The Journey is Just as Important as the Destination

Warnings and Hazards

The risks and hazards during this second experiment are minor however any time that you are in a lab, you must be cautious. The best way to remain safe is to understand RAMP.

Recognize hazards,
Assess risks,
Minimize risks, and
Prepare for emergencies

Using the space below, comment on each of the portions of the RAMP analysis below.

Recognize hazards (what are potential hazards that will be encountered in the procedure):
_____
_____
_____
_____
_____

Assess risks (how can these hazards be harmful to people or your environment):
_____
_____
_____
_____
_____

Minimize risks (what can you do to limit risk yourself and your environment):
_____
_____
_____
_____
_____

Prepare for emergencies.
Emergency: 911
Stockroom: 512-245-3118

Background

Copper is represented on the periodic table by the symbol "Cu" and has been used by humans for thousands of years. In addition to being of use as a pure solid, copper is of use to living organisms as an ion or as a part of a compound. Copper generally exists as a +2 ion, the second most common copper ion is a +1 ion BUT copper can also exist with a charge of -2, 0, +3, and +4.

In this experiment we will explore some of the different reactions that copper goes through. With these different experiments we will be looking at how copper can form soluble and insoluble compounds. Through this experiment we will be exploring aqueous substances and solid substances, molecular, ionic, and net ionic reactions, as well as oxidation numbers. Although it may seem like we are simply looking at copper and how it changes, we are going to be relating these changes that copper undergoes to properties of solutions and different terminologies that we will use to describe substances.

In addition to the transitions that copper will undergo, we will investigate the basis for a medical test known as the *Benedict's test*. This test was developed to detect glucose in urine, which is a condition known as glucosuria. If someone exhibits glucosuria that is an indication that they may have diabetes mellitus. This test is all related to copper and the ions it can form and how those ions react with glucose.

Experimental Objective

The point of this experiment is to observe how one cation, copper reacts with different anions to form different insoluble compounds. This experiment will also help you understand and visualize differences between insoluble and soluble compounds. We will also be able to investigate single and double displacement reactions as well as oxidation and reduction.

Bin Contents

All of your glassware will be in your bin and that bin will be collected from the stockroom after you have successfully filled out a copy of the glassware pickup document found at the back of your lab manual.

In the space below, fill in the blanks below to have a complete list of your week's glassware. In the space on the right of this page, draw each piece of glassware.

Glassware	Quantity	Capacity
_____	_____	
_____	_____	
_____	_____	
_____	_____	
_____	_____	
_____	_____	
_____	_____	
_____	_____	
_____	_____	
_____	_____	
_____	_____	
_____	_____	

Procedure
Part I:
Oxidizing Copper Metal with Concentrated Nitric Acid

Note: this portion of your experiment will be performed under the fume hood, the nitric acid will be added to your solid copper in a dropwise fashion

Note: it is absolutely essential that you wear gloves for this procedure!

1. Determine the mass of your glass beaker and record it in the space to the left. Label this glass beaker with your initials.
2. Determine the mass of your solid copper and record it in the space to the left and place the copper in your glass beaker from step 1.
3. Using the glass dropper under the fume hood, carefully transfer nitric acid to your solid copper - as soon as the copper has completely dissolved, stop adding the acid to that beaker. Record the number of drops of nitric acid that were added to the copper in the space to the left.
4. Leave your glass beaker in the fume hood for approximately five minutes.
5. Record any observations that you may have in the space provided on the left - was a gas produced? did this reaction happen rapidly? were there any color changes?
6. After recording any observations, bring your beaker to your workstation.
7. Add approximately 10mL of deionized water to your copper containing mixture.
8. Record what is in your beaker at this point in time in the space to the left - think about what atoms and ions are present.

Addition of HNO_3 to solid copper

Mass of glass beaker _____ g
Mass of solid copper _____ g

Drops of HNO_3 _____

Charge of <u>solid</u> copper _____

Identity of gas produced _____

Comments: _____
_____
_____
_____
_____
_____
_____

What is the charge of the copper that is in your beaker? (hint it has changed)
_____

What polyatomic ions are present in your beaker?
_____

What else is in your beaker? _____
_____
_____
_____

Part II:
Neutralizing your solution

Note: it is absolutely essential that you wear gloves for this procedure!

1. Carefully place a magnetic stirrer in your beaker from part I.
2. Place your beaker from part I onto your hot plate/stir plate, adjust the "Stir" dial to between 60 & 200. Be careful to **not** adjust the "Heat" dial.
3. Dip your glass stir rod into the solution in your beaker and transfer a small amount of the liquid from your beaker onto your pH paper.
4. Begin adding sodium hydroxide, drop by drop, to your beaker and watch as the solution changes, record your observations in the space to the right.
5. After transferring 5 drops, again use the glass stir rod to check the pH of your solution on the pH paper.
6. Continue transferring NaOH, 5 drops at a time, and checking the pH until your solution is basic.
7. Add approximately 10mL of deionized water to your copper containing mixture.
8. Record what is in your beaker at this point in time in the space to the left - think about what atoms and ions are present.

Part III:
Generating copper (II) oxide

1. Continue stirring the solution but now adjust the heat of your hot plate - turn your dial to ~10 o'clock (keep in mind this dial does not have number markings, look at it like the face of an analog clock and adjust to 9 or 10).
2. Let this mixture stir while heating, this reaction will be complete when all of the solid has changed color.
3. As soon as this solid has changed color turn the heat and stir function on your hot plate off.
4. Transfer your beaker to your hot pad and allow it to cool for 5 minutes

---

Addition of NaOH
What color is your pH paper when the solution is basic?
_____

Observations: _____
_____
_____
_____
_____
_____
_____
_____

What is the metathesis reaction for what has mixed in your solution?
(hint: think about the ions that were in and the ions that you have added)
_____
_____

What color is the solid after heating?
_____

What is the solid after heating?
_____

5. Place your glass funnel in the top of your erlenmeyer flask.
6. Remove the baffles from your coffee filter and place it into your glass funnel.
7. Transfer the contents of your beaker into the coffee filter, being sure to transfer all of the solid.
8. Wash the solid in the coffee filter with approximately 20-30mL of deionized water.
9. Transfer your glass funnel to your other erlenmeyer flask.
10. Discard the solution in your first erlenmeyer flask.

Part IV:
Recovering Copper ions
1. Carefully and slowly add approximately 10mL of sulfuric acid to your coffee filter + funnel + copper oxide.
2. If needed add more sulfuric acid to react with any remaining copper oxide.
3. The copper ions should form a faint blue solution in the erlenmeyer flask.
4. Carefully transfer the solution from the erlenmeyer flask to your beaker.
5. Place a magnetic stirrer in your beaker.

Part V:
Recovering Solid Copper
1. Fill in the flow chart to the left (include states like (aq) and/or (s)).
2. In the space below, use dimensional analysis to show the relationship between the initial amount of copper used and the amount of Zn needed.

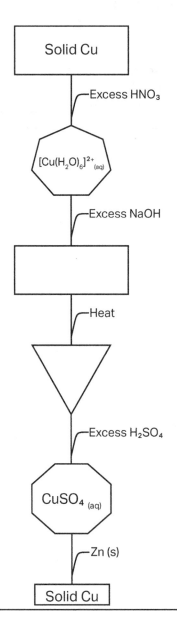

3. Based on your dimensional analysis, record the mass of Zn needed in the space to the right.
4. Carefully measure out the approximate mass needed - record the exact mass that you got in the space to the right.
5. Place your beaker from part IV with the magnetic stirrer on your stir/hot plate - adjust the "Stir" dial to between 60 and 200 - do NOT heat your sample.
6. Carefully transfer the solid Zn into your beaker with the magnetic stirrer.
7. If your solution has not complete lost the blue hue with the addition of the Zn - add a small amount of additional Zn (no more than 50% of what was previously added) - record the exact mass of what was added.
8. As soon as your solution has lost all of the blue hue, collect 10mL of H_2SO_4 and add it to your beaker.
9. Remove and discard the baffles from a coffee filter - determine the dry mass of your coffee filter.
10. Place your funnel in your erlenemeyer flask, place the debaffled coffee filter in the funnel and carefully transfer the contents of your beaker into the coffee filter.
11. Ensure that all solids have been transferred into the coffee filter by rinsing your beaker with deionized water - retrieve your magnetic stirrer.
12. Determine the mass of your watch glass.
13. Carefully transfer your coffee filter to your pre-weighed watch glass and place them into the oven.
14. Incubate your sample in the oven for 10 minutes, with a hot pad, carefully remove your watch glass, coffee filter, and sample from the oven and determine the mass.
15. To determine the mass of the solid Cu, subtract the mass of the CF and WG from the combined mass of the solid Cu, CF, and WG.
16. The percent Cu recovered is calculated by dividing mass of Cu from part 5 by the mass of Cu from part 1 and multiplying that quotient by 100 (to make it a percentage).

Before drying
Mass of Zn required (Zn) _____
Mass of coffee filter (CF) _____
Mass of watch glass (WG) _____

After 10min drying
Mass of solid + CF + WG _____
Mass of solid Cu _____

Mass of solid Cu from pt 1 _____
Mass of Cu difference _____
Percent Cu recovered _____

Observations

Part VI:
Copper with different oxidation numbers?

1. With a clean beaker, collect 25mL of the copper hydroxide solution.
2. Add a magnetic stirrer to the copper hydroxide solution and adjust "stir" dial to between 60-200.
3. Collect 1g of glucose and add it to the beaker on the stir plate.
4. Adjust the dial on your "heat" to the 10 o'clock position.
5. This solution will begin to change similar to how your previous copper hydroxide solution changed when heated.
6. Record your observations of what happens in the space to the left.
7. After the solid in your reaction has completed changed color - turn off the stir & heat functions of your hot plate. Discard the resulting solution, be certain to recover your stir bars.

Reflections

Answer the questions below using your data - critically evaluate and analyze your results when answering these questions.

1. In part I of the procedure, what color is the gas that was produced?

2. What reactants are present in part I?

3. Considering the reactants present in part I - what elements are present in the gas that was generated?

4. Describe the difference between a substance that is a solid and identified with the subscript (s) compared to something identified with the subscript (aq) meaning aqueous. Think about these compounds on the atomic level - how are they different from one another?

5. Part I of our experiment shows a strong acid, HNO_3 reacting with copper - HCl is also a strong acid but will not react with copper. Hydrogen ions will not react with Cu - considering this information and these two strong acids, what ions react with Cu?

6. Considering your answer to question 3 and 5 - what do you think is the identity of the gas generated in part I?

7. In part III of your procedure, what is the formula for the copper containing compound before the heating and what is the formula for the copper containing compound after heating?

8. In part III and part VI of your procedure, you make copper oxide solids - the only difference between them is that in part III your make copper (II) oxide and in part VI you make copper (VI) oxide - considering the solid that you started each of these parts with, is copper oxidized or reduced in either? If so - what takes place in each part?

9. In parts III and VI - is heat responsible for the oxidation/reduction?

10. Review the activity series shown on the next page - what other metals could have been used instead of zinc - could any other metals be used?

11. In part V of your procedure, H_2SO_4 was used - could an acid like HCl be used for this procedure? What about HNO_3?

12. What are the charges of the different copper ions that are encountered in this experiment?

13. If you form Cu_2O in part VI - what is the charge of the copper ion in that compound?

Activity Series of Metals

Metal	Oxidation Half-Reaction
Lithium	$Li_{(s)} \rightarrow Li^+_{(aq)} + e^-$
Potassium	$K_{(s)} \rightarrow K^+_{(aq)} + e^-$
Barium	$Ba_{(s)} \rightarrow Ba^{2+}_{(aq)} + 2e^-$
Calcium	$Ca_{(s)} \rightarrow Ca^{2+}_{(aq)} + 2e^-$
Sodium	$Na_{(s)} \rightarrow Na^+_{(aq)} + e^-$
Magnesium	$Mg_{(s)} \rightarrow Mg^{2+}_{(aq)} + 2e^-$
Aluminum	$Al_{(s)} \rightarrow Al^{3+}_{(aq)} + 3e^-$
Manganese	$Mn_{(s)} \rightarrow Mn^{2+}_{(aq)} + 2e^-$
Zinc	$Zn_{(s)} \rightarrow Zn^{2+}_{(aq)} + 2e^-$
Chromium	$Cr_{(s)} \rightarrow Cr^{3+}_{(aq)} + 3e^-$
Iron	$Fe_{(s)} \rightarrow Fe^{2+}_{(aq)} + 2e^-$
Cobalt	$Co_{(s)} \rightarrow Co^{2+}_{(aq)} + 2e^-$
Nickel	$Ni_{(s)} \rightarrow Ni^{2+}_{(aq)} + 2e^-$
Tin	$Sn_{(s)} \rightarrow Sn^{2+}_{(aq)} + 2e^-$
Lead	$Pb_{(s)} \rightarrow Pb^{2+}_{(aq)} + 2e^-$
Hydrogen	$H_{2(g)} \rightarrow 2H^+_{(aq)} + 2e^-$
Copper	$Cu_{(s)} \rightarrow Cu^{2+}_{(aq)} + 2e^-$
Silver	$Ag_{(s)} \rightarrow Ag^+_{(aq)} + e^-$
Mercury	$Hg_{(s)} \rightarrow Hg^{2+}_{(aq)} + 2e^-$
Platinum	$Pt_{(s)} \rightarrow Pt^{2+}_{(aq)} + 2e^-$
Gold	$Au_{(s)} \rightarrow Au^{3+}_{(aq)} + 3e^-$

Look to the chemical equation - the metal on the left will be oxidized when in the presence of ions on the right IF the metal is above those ions on the activity series.

Mixing zinc metal with copper ions will result in electrons being transferred from zinc to the copper ions, forming copper metal and zinc ions.

Mixing zinc ions with copper ions will do nothing.

Mixing nickel metal with cobalt ions will do nothing. However mixing cobalt metal with nickel ions will produce cobalt ions and nickel metal.

Mixing sodium with an acid produces H_2 gas - will mixing silver with an acid also produce H_2 gas?

Safety Data Sheets

The QR code shown below will take you to the Safety Data Sheet for the chemical(s) used in this experiment

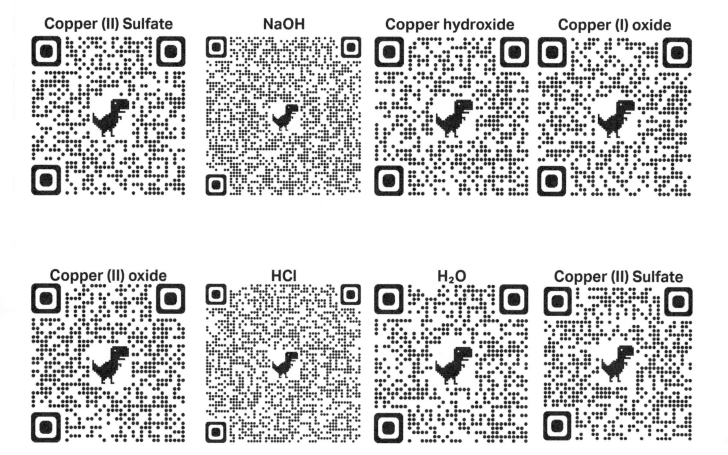

CHEMISTRY 1141
GENERAL CHEMISTRY LABORATORY

Experiment #5
How Much is there?
A Simple Acid-Base Titration

Warnings and Hazards

The risks and hazards during this second experiment are minor however any time that you are in a lab, you must be cautious. The best way to remain safe is to understand RAMP.

Recognize hazards,
Assess risks,
Minimize risks, and
Prepare for emergencies

Using the space below, comment on each of the portions of the RAMP analysis below.

Recognize hazards (what are potential hazards that will be encountered in the procedure):
_____
_____
_____
_____

Assess risks (how can these hazards be harmful to people or your environment):
_____
_____
_____
_____

Minimize risks (what can you do to limit risk yourself and your environment):
_____
_____
_____
_____

Prepare for emergencies.
Emergency: 911
Stockroom: 512-245-3118

Background

The purpose of this experiment is to help you understand how to use a specialized piece of glassware known as a buret. The other purpose for this experiment is for you to learn how to do a simple acid-base titrtation. As a technique, a titration is an invaluable means of determining the concentration of a substance, given knowledge of the concentration of another substance that reacts with the substance of unknown concentration. This is an acid-base titration which is one of the more common titrations done in a General Chemistry lab, but it is a great learning tool to understand the exact quantity and amount of an base that is required to neutralize a acid or vice-versa.

Titrations do not simply need to revolve around acids and bases but can be used for any type of reaction in which you can use an indicator solution. An indicator solution will simply show when the reaction has stopped. Indicator solutions have different properties but generally show a color change when a certain pH has been attained.

An example of the use of titrations outside of a chemistry teaching lab would be the frequent use of titrations in a clinical lab to detect certain compounds in blood or urine.

Experimental Objective

The point of this experiment is to understand how acids and bases react. This experiment will also show how to use a specific piece of glassware, a buret. We will also gain familiarity to a commonly used tool in a chemistry lab - an indicator solution.

Bin Contents

All of your glassware will be in your bin and that bin will be collected from the stockroom after you have successfully filled out a copy of the glassware pickup document found at the back of your lab manual.

In the space below, fill in the blanks below to have a complete list of your week's glassware. In the space on the right of this page, draw each piece of glassware.

Glassware	Quantity	Capacity
_____	_____	_____
_____	_____	_____
_____	_____	_____
_____	_____	_____
_____	_____	_____
_____	_____	_____
_____	_____	_____
_____	_____	_____
_____	_____	_____
_____	_____	_____
_____	_____	_____

Procedure
Part I:
Equilibrating and preparing your buret

Note: you will be handling the strong base, sodium hydroxide, NaOH - it is imperative that you wear gloves when handling this solution

1. After collecting your buret, position it in the proper clamp at your lab work station.
2. The valve of the buret should be parallel to your bench top and it should be perpendicular to your buret.
3. In your plastic beaker, collect approximately 50mL of deionized water.
4. Place your funnel in the opening at the top of your buret and carefully transfer approximately 5mL of deionized water into the buret.
5. With a gloved hand, grab the top of the buret and place your thumb over the opening at the top of the buret and invert the buret several times - your objective is to coat the interior of the buret with deionized water.
6. After washing the interior of the buret has been completely coated with deionized water, place the buret back in the stand and adjust the valve to be completely perpendicular with the benchtop and let all the water flow out.
7. Collect the water that flowed out of the buret and discard it.
8. Now collect approximately 60mL of sodium hydroxide - record the concentration of the solution in the space to the left.
9. Place your funnel in the top of the buret and transfer approximately 5-10mL of sodium hydroxide to the buret.
10. Repeat the same rinsing of the interior of the buret but now with NaOH.
11. Elute the buret contents and discard that solution.
12. Add NaOH into the buret so the meniscus is exactly at 0mL.

Data
Concentration of NaOH _____

Part II:

Preparing your unknown solution

1. Collect exactly 25mL of the HCl solution (concentration is unknown) using your graduated cylinder and transfer that to your erlenmeyer flask.
2. Collect exactly 25mL of deionized water and transfer that to your erlenmeyer flask.
3. Transfer approximately 3-5 drops of your indicator solution into the erlenmeyer flask containing the unknown acid solution.
4. Place your erlenmeyer flask with the unknown acid underneath your buret and open the valve on the buret to transfer sodium hydroxide.
5. Transfer a small amount of the sodium hydroxide into the acid, elute approximately 0.5mL at a time while gently swirling the erlenmeyer flask with the acid solution.
6. Your acid solution will gradually change color from a clear solution to a light pink solution and then to a solid purple solution. You do not want a solid purple solution - you want a very faint pink solution.
7. As soon as your solution is a faint pink color, stop transferring the base from the buret into the erlenmeyer flask and record the volume of base transferred in the space provided to the right.
8. Discard the solution in your erlenmeyer flask in the proper waste container and rinse your erlenmeyer flask with deionized water.
9. Repeat steps 1-8 for trials 2 & 3 - recording the data for those trials on the next page.

Trial 1

Volume of HCl solution _____ mL
Volume of HCl solution _____ L
Drops of phenolphthalein _____

Inital buret reading _____
Final buret reading _____
Volume of NaOH used _____ mL
Volume of NaOH used _____ L
Molarity of NaOH _____ M
Moles of NaOH used _____ mol

__HCl + __NaOH -> __NaCl + __H$_2$O

Molar ratio of NaOH:HCl _____
Moles of HCl titrated _____ mol
Molarity of HCl solution _____ M

Trial 2
Volume of HCl solution _____ mL
Volume of HCl solution _____ L
Drops of phenolphthalein _____

Inital buret reading _____
Final buret reading _____
Volume of NaOH used _____ mL
Volume of NaOH used _____ L
Molarity of NaOH _____ M
Moles of NaOH used _____ mol

__HCl + __NaOH -> __NaCl + __H$_2$O

Molar ratio of NaOH:HCl _____
Moles of HCl titrated _____ mol
Molarity of HCl solution _____ M

Trial 3
Volume of HCl solution _____ mL
Volume of HCl solution _____ L
Drops of phenolphthalein _____

Inital buret reading _____
Final buret reading _____
Volume of NaOH used _____ mL
Volume of NaOH used _____ L
Molarity of NaOH _____ M
Moles of NaOH used _____ mol

__HCl + __NaOH -> __NaCl + __H$_2$O

Molar ratio of NaOH:HCl _____
Moles of HCl titrated _____ mol
Molarity of HCl solution _____ M

Collective Data
Molarity of HCl solution from Trial 1 _____ M
Molarity of HCl solution from Trial 2 _____ M
Molarity of HCl solution from Trial 3 _____ M

Average Molarity _____ M
Which trial is closest to the calculated average? _____
Which trial is farthest from the calculated average? _____

Reflections

Answer the questions below using your data - critically evaluate and analyze your results when answering these questions.

1. HCl is considered a strong acid because it will completely dissociate in solution - what does that mean?

2. NaOH is considered a strokg base because it will completely dissociate in solution - what does that mean?

3. Based on your results and observations, if you were to add phenolphthalein directly to a sodium hydroxide solution, what color would that solution be?

4. If you were to complete a fourth trial, what volume of NaOH would you expect to use?

5. How would your results differ if you used a NaOH solution with a higher concentration?

6. If HCl is a strong acid and will completely dissociate in solution, what do you think the definition is for a weak acid?

7. How does a weak acid differ from a strong acid on the molecular level?

8. What does phenolphthalein tell you about a solution (hint: is the solution acidic, basic, or neutral)

Safety Data Sheets
The QR code shown below will take you to the Safety Data Sheet for the chemical(s) used in this experiment

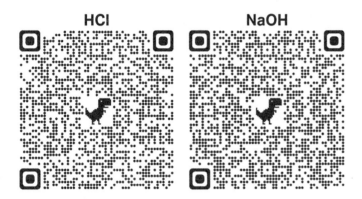

CHEMISTRY 1141
GENERAL CHEMISTRY LABORATORY

Experiment #6
Titration² The Remix

Warnings and Hazards

The risks and hazards during this second experiment are minor however any time that you are in a lab, you must be cautious. The best way to remain safe is to understand RAMP.

Recognize hazards,
Assess risks,
Minimize risks, and
Prepare for emergencies

Using the space below, comment on each of the portions of the RAMP analysis below.

Recognize hazards (what are potential hazards that will be encountered in the procedure):
_____
_____
_____
_____

Assess risks (how can these hazards be harmful to people or your environment):
_____
_____
_____
_____
_____

Minimize risks (what can you do to limit risk yourself and your environment):
_____
_____
_____
_____

Prepare for emergencies.
  Emergency: 911
  Stockroom: 512-245-3118

Background

This experiment adds a bit more complexity to our previous titration. This experiment has multiple layers because, as you would imagine, most solutions that you work with on a day to day basis are not as simple as having one acid - let alone that one acid is a strong acid. To address and investigate this, we will do a titration with a dilute orange juice solution.

Orange juice has two substances that you are no doubt familiar with - vitamin C and citric acid. Both of these are classified as weak acids. For the remainder of this procedure we will refer to vitamin C by its more techical name, ascorbic acid. Similar to what we did last week, we can use sodium hydroxide to titrate these acids. Titrating with sodium hydroxide will help us figure out the total amount of acid in the solution. We will have to take the procedure to a new level to determine the specific amounts of ascorbic and citric acid in this solution.

A simple way to look at this experiment is to break it into three different parts as described below and detailed on the next pages.

Part I: Determine total amount of acid in solution

Part II: Determine amount of ascorbic acid in solution

Part III: Data analysis/

Procedure
Part I:
Determination of total amount of acid

Arguably, this portion of the procedure is most similar to what you did in last week's experiment. Last week we worked with the strong acid, HCl which is also known as a monoprotic acid due to the fact that the molecule only has one proton to be neutralized. This week we will be working with the weak acids cirtic acid and ascorbic acid. Ascorbic acid is a monoprotic acid and citric acid is a triprotic acid (shown to the right). The protons that each weak acid will lose are underlined. Take a moment to label the molecules to the right and balance the chemical reaction where they react with sodium hydroxide.

Name _____
Molecular Formula _____

____ + __ NaOH -> ____ + ____

Part II:
Determination of ascorbic acid concentration

This second titration seems quite a bit different but keep in mind what the differences are. Rather than using a strong base like sodium hydroxide in your buret, you will be using an ionic solution. Your ionic solution is potassium iodate which will produce I_2 when in an acidic environment. The resulting I_2 will indicate that ascorbic acid will be oxidized - another way of saying that I_2 reduces ascorbic acid.

In part I of the procedure your solution will be slightly pink when enough of the strong base has been added. In this part of the procedure, your solution will be a faint navy blue color. That blue solution is the result of the <u>iodine (I_2) interacting with the starch</u> solution.

Name _____
Molecular Formula _____

Part III:
Data Analysis

Arguably, this portion of the procedure is the most important porition considering that without this, you cannot really say anything definitive about the work done in parts 1 & 2. All that you would be able to say for parts 1 & 2 is something along the

____ + __ NaOH -> ____ + ____

lines of "one solution turned a faint pink color and the other solution turned a faint blue color!"

We want to analyze the mixtures of solutions that helped us attain those faint blue and pink solutions. The calculations for part 1 are highly similar to what we saw last week due to the similarities to the techniques. For example, last week we had a balanced chemical reaction and used that chemical reaction to look determine the concentration of acid. With that in mind, it is important to remember that last week we used hydrochloric acid (HCl) which is a monoprotic acid and sodium hydroxide will dissociate to make a single hydroxide ion. HCl and NaOH pair up in a 1:1 ratio - if you have 1 mole of NaOH then 1 mole of HCl was present in your solution. It was a somewhat simple and straightforward situation!

In contrast to the simple and straightforward situation with NaOH and HCl, we're working with two different acids! The acids that we are working with are monoprotic and triprotic as well! Despite this difference, this doesn't radically change the math and what we need to do. A triprotic acid (top left) and a monoprotic acid (bottom left), for all intents and purposes, are equivalent to a tetraprotic acid. Sodium hydroxide and the acids basically pair up in a 1:4 ratio!

The final calculations are only more complicated because of the players involved - the iodate ion and ascorbic acid. At the end of the day, it is all just a ratio and as it turns out, our ratio in this portion of the experiment is a 1:3:1:1 ratio. For every mole of iodate ion used, 3 moles of I_2 will be produced, and for every 1 mole of I_2 produced, 1 mole of ascorbic acid will be oxidized. This all then boils down to 1 mole of iodate ion will oxidize 1 mole of ascorbic acid - albeit in a somewhat scattered fashion.

Experimental Objective

The point of this experiment is to understand how acids and bases react. This experiment will also show how to use a specific piece of glassware, a buret. We will also gain familiarity to a commonly used tool in a chemistry lab - an indicator solution.

Bin Contents

All of your glassware will be in your bin and that bin will be collected from the stockroom after you have successfully filled out a copy of the glassware pickup document found at the back of your lab manual.

In the space below, fill in the blanks below to have a complete list of your week's glassware. In the space on the right of this page, draw each piece of glassware.

Glassware	Quantity	Capacity
_____	_____	_____
_____	_____	_____
_____	_____	_____
_____	_____	_____
_____	_____	_____
_____	_____	_____
_____	_____	_____
_____	_____	_____
_____	_____	_____
_____	_____	_____
_____	_____	_____

Procedure
Part 0:
Equilibrating and preparing your buret

Note: you will be handling the strong base, sodium hydroxide, NaOH - it is imperative that you wear gloves when handling this solution

1. After collecting your buret, position it in the proper clamp at your lab work station.
2. The valve of the buret should be parallel to your bench top and it should be perpendicular to your buret.
3. In your plastic beaker, collect approximately 50mL of deionized water.
4. Place your funnel in the opening at the top of your buret and carefully transfer approximately 5mL of deionized water into the buret.
5. With a gloved hand, grab the top of the buret and place your thumb over the opening at the top of the buret and invert the buret several times - your objective is to coat the interior of the buret with deionized water.
6. After washing the interior of the buret has been completely coated with deionized water, place the buret back in the stand and adjust the valve to be completely perpendicular with the benchtop and let all the water flow out.
7. Collect the water that flowed out of the buret and discard it.
8. Now collect approximately 60mL of sodium hydroxide - record the concentration of the solution in the space to the left.
9. Place your funnel in the top of the buret and transfer approximately 5-10mL of sodium hydroxide to the buret.
10. Repeat the same rinsing of the interior of the buret but now with NaOH.
11. Elute the buret contents and discard that solution.
12. Add NaOH into the buret so the meniscus is exactly at 0mL.

Part I:

Determining the amount of acid in fruit juice

1. Collect exactly 25mL of the fruit juice solution in your graduated cylinder and add it to your erlenmeyer flask.
2. Collect exactly 25mL of deionized water and transfer that to your erlenmeyer flask.
3. Transfer approximately 3-5 drops of your indicator solution into the erlenmeyer flask containing the unknown acid solution.
4. Place your erlenmeyer flask with the fruit juice solution underneath your buret.
5. Transfer a small amount of the sodium hydroxide into the acid, elute approximately 0.5mL and gently swirl the erlenmeyer flask with the acid solution.
6. Continue transferring small amounts of sodium hydroxide to the acid - transfer approximately 0.5mL at a time.
7. Watch as your juice solution will gradually change colors to a light pink solution and if you're not careful, to a solid purple solution. <u>You do not want a solid purple solution - you want a steady but very faint pink solution.</u>
8. As soon as your solution is a faint pink color, stop transferring the base from the buret into the erlenmeyer flask and record the volume of base transferred in the space provided to the right.
9. Discard the solution in your erlenmeyer flask in the proper waste container and rinse your erlenmeyer flask (3x) with deionized water.
10. Repeat steps 1-9 for trials 2 & 3 - recording the data for those trials on the next page.
11. If necessary you will need to fill your buret with additional NaOH until it reaches zero. Record the exact reading under the "initial buret reading" on your data sheet under trial 2. It is OK if the NaOH level is not at exactly zero, but what is critical is that you write the exact reading of the buret.
12. As soon as you have finised the titration for trial 3, complete the following.
    -dispose of the solution in the Erlenmeyer flask in the proper waste container

Trial 1
Volume of juice solution _____ mL
Volume of juice solution _____ L
Drops of phenolphthalein _____

Inital buret reading _____
Final buret reading _____
Volume of NaOH used _____ mL
Volume of NaOH used _____ L
Molarity of NaOH _____ M
Moles of NaOH used _____ mol

Trial 2
Volume of juice solution _____ mL
Volume of juice solution _____ L
Drops of phenolphthalein _____

Inital buret reading _____
Final buret reading _____
Volume of NaOH used _____ mL
Volume of NaOH used _____ L
Molarity of NaOH _____ M
Moles of NaOH used _____ mol

Trial 3
Volume of juice solution _____ mL
Volume of juice solution _____ L
Drops of phenolphthalein _____

Inital buret reading _____
Final buret reading _____
Volume of NaOH used _____ mL
Volume of NaOH used _____ L
Molarity of NaOH _____ M
Moles of NaOH used _____ mol

Trial 1
Volume of juice solution	_____ mL
Volume of juice solution	_____ L
Drops of phenolphthalein	_____
Inital buret reading	_____
Final buret reading	_____
Volume of KIO₃ used	_____ mL
Volume of KIO₃ used	_____ L
Molarity of KIO₃	_____ M
Moles of KIO₃ used	_____ mol

Trial 2
Volume of juice solution	_____ mL
Volume of juice solution	_____ L
Drops of phenolphthalein	_____
Inital buret reading	_____
Final buret reading	_____
Volume of KIO₃ used	_____ mL
Volume of KIO₃ used	_____ L
Molarity of KIO₃	_____ M
Moles of KIO₃ used	_____ mol

Trial 3
Volume of juice solution	_____ mL
Volume of juice solution	_____ L
Drops of phenolphthalein	_____
Inital buret reading	_____
Final buret reading	_____
Volume of KIO₃ used	_____ mL
Volume of KIO₃ used	_____ L
Molarity of KIO₃	_____ M
Moles of KIO₃ used	_____ mol

-carefully pour any remaining NaOH in the buret into a waste beaker. Dispose of the contents of your waste beaker in the proper waste container
-wash your buret with soap and water (use diluted soap in squirt bottle), rinse thoroughly and lightly spritz the inside of the buret with ethanol from the ethanol squirt bottle and drain
-rinse all glassware with deionized H_2O
-dry glassware and return to bin after rinse

Part II:
Determine the amount of ascorbic acid in solution
1. Collect exactly 25mL of the fruit juice solution in your graduated cylinder and add it to your erlenmeyer flask.
2. Collect 25mL of deionized water to your erlenmeyer flask.
3. Add 10 drops of the 3% starch solution to your erlenmeyer flask.
4. Using a transfer pipet, add 1mL of HCl to your erlenmeyer flask.
5. Similar to part 0, you will have to equilibrate your buret however you will not be using sodium hydroxide to equilibrate the buret but rather you will be using potassium iodate.
6. After equilibrating your buret with the potassium iodate solution, add potassium iodate to the buret to the 0mL marking.
7. Transfer a small volume of potassium iodate to the erlenmeyer flask and swirl the solution to ensure that it is mixed.
8. Begin transferring small amounts of potassium iodate at a time, for example, transfer 0.5mL and watch for a color change.
9. Continue transferring (extremely gradually) potassium iodate to the erlenmeyer flask until a faint blue color is observed.
10. As soon as the color change is permanent, stop transferring and record the final volume.

11. Record the final volume of potassium iodate transferred to the erlenmeyer flask.
12. Record other necessary data in the space provided on the previous page.
13. Repeat this titration two more times and record data from those additional titrations in the spaces provided on the previous page.

Part III:
Processing the Data
1. Calculate the average number of moles of NaOH that were required for part I and input that value in the space to the right.
2. Calculate the number of acidic protons that need to be neutralized between ascorbic acid and citric acid - enter this information in the space to the right (see background information).
3. Calculate the molar ratio of NaOH to acidic protons and enter that information in the space to the right (see background balanced chemical reaction).
4. Calculate the total number of moles of acidic protons in the fruit juice sample and add to space to the right.

Calculations

Average moles of NaOH _____

Acidic protons to neutralize _____

Molar ratio _____

Total moles of acidic protons _____

BREAK

5. Calculate the average number of moles of KIO_3 from part 2 and write that number in the space on the right.
6. Using the dimensional analysis displayed below, calculate the number of moles of ascorbic acid in fruit juice.

Calculations

Average moles of KIO_3 _____

Moles of ascorbic acid _____

Moles of acidic protons _____

$$\frac{\underline{} \text{ mole } KIO_3}{} \times \frac{3 \text{ mole } I_2}{1 \text{ mole } KIO_3} \times \frac{1 \text{ mole ascorbic acid}}{1 \text{ mole } I_2} = \underline{} \text{ mole ascorbic acid}$$

7. Using the dimensional analysis below, calculate the number of moles of acidic protons from ascorbic acid.

$$\frac{\underline{} \text{ mole ascorbic acid}}{} \times \frac{1 \text{ mole acidic protons}}{1 \text{ mole ascorbic acid}} = \underline{} \text{ mole acidic protons from ascorbic acid}$$

8. Subtract the number of moles of acidic protons from ascorbic acid from the total number of moles of acidic protons - that is the number of moles of acidic protons from citric acid.

9. Using the dimensional analysis below, calculate the number of moles of citric acid in the juice solution.

$$\frac{\_\_\_ \text{ mole of acidic protons from citric acid}}{} \times \frac{1 \text{ mole citric acid}}{3 \text{ mole acidic protons from citric acid}} = \_\_\_ \text{ mole of citric acid}$$

Reflections

Answer the questions below using your data - critically evaluate and analyze your results when answering these questions.

1. How many moles of total acidic protons were in the juice solution?

2. How many moles of ascorbic acid were in the juice solution?

3. How many moles of citric acid were in the juice solution?

4. What acid was more concentrated in the juice solution?

5. Would your results vary if instead of 1 monoprotic acid and one triprotic acid you had two diprotic acids?

6. How would your results vary if you had two monoprotic acids?

7. In our titration, ascorbic acid reacts with I_2 and I^- is a product of this reaction - has I_2 been reduced or oxidized by ascorbic acid?

8. If you have a diet high in shrimp and other seafood, you're at a greater risk of iodine poisoning - would taking a citric acid supplement or vitamin C supplement be better for you? Explain.

CHEMISTRY 1141
GENERAL CHEMISTRY LABORATORY

Experiment #7
Metal Reactivities
Single & Double Displacement Reactions

Warnings and Hazards

The risks and hazards during this second experiment are minor however any time that you are in a lab, you must be cautious. The best way to remain safe is to understand RAMP.

Recognize hazards,
Assess risks,
Minimize risks, and
Prepare for emergencies

Using the space below, comment on each of the portions of the RAMP analysis below.

Recognize hazards (what are potential hazards that will be encountered in the procedure):
_____
_____
_____
_____

Assess risks (how can these hazards be harmful to people or your environment):
_____
_____
_____
_____

Minimize risks (what can you do to limit risk yourself and your environment):
_____
_____
_____
_____

Prepare for emergencies.
Emergency: 911
Stockroom: 512-245-3118

Background

In this experiment you will be investigating some fairly simple and straightforward reactions. Your primary objective is to understand how a solid metal, a material in its elemental state will react with a solution containing ionic compounds. You'll also be investigating the opposite with the goal to understand why a metal does NOT react with a solution containing ions.

In addition to investigating how metals react with ionic solutions, we will also be investigating how ionic solutions react with one another. A reaction taking place between two ionic solutions will be indicated by the formation of an insoluble solid compound. This experiment will help you gain a stronger understanding of several subjects discussed in your lecture - a solubility table, metal activity series, as well as single and double displacement reactions. These topics will all be briefly addressed below.

Solubility: if a substance dissolves and dissociates in solution, that substance is said to be soluble - the solution will typically be transparent (though some may have a color) and in a chemical reaction will be signified by (aq)

Metal activity series: if a metal appears at the top of the metal activity series table it will more easily be oxidized and is more reactive than a metal at the bottom of the table

Single & Double Displacement Reactions:

Single: A + BX -> AX + B

Double: AX + BY -> BX + AY

Experimental Objective

The point of this experiment is to gain a greater understanding of solution chemistry as well as how metals react with different solutions.

Bin Contents

All of your glassware will be in your bin and that bin will be collected from the stockroom after you have successfully filled out a copy of the glassware pickup document found at the back of your lab manual.

In the space below, fill in the blanks below to have a complete list of your week's glassware. In the space on the right of this page, draw each piece of glassware.

Glassware	Quantity	Capacity
_____	_____	_____
_____	_____	_____
_____	_____	_____
_____	_____	_____
_____	_____	_____
_____	_____	_____
_____	_____	_____
_____	_____	_____
_____	_____	_____
_____	_____	_____
_____	_____	_____
_____	_____	_____

Procedure
Part 1:
Evaluating reactions between metals and solutions

1. From the fume hood, collect a well plate, familiarize yourself with the grid/matrix of the well plate.
2. Collect small amounts of each of the unknown metals that you will be working with and bring them back to your workspace.
3. Collect an index card which displays the grid and matrix of solutions and metals to combine.
4. For each of the ionic solutions that you will be using provide a brief description of them in the space to the left. Does the solution have a color? Is there anything identifying about that solution other than the label? What is the concentration of the solution?
5. Transfer approximately 1-2mL of each ionic solution to the corresponding wells on the well plate.
6. In the space provided below, write your observations about the unknown metals (color, shiny, etc) and make a prediction about the identity - this may not be accurate but take a chance, yolo.
7. Transfer small amounts of the unknown metals to the corresponding wells on the well plate.
8. Allow this plate to sit and incubate with the lid on for the next 30 minutes - after that, use the activity series (pg 72) and your knowledge of the solutions and metals on the next page to revise your prediction.

Ionic Solutions

Solution #1 - _____
_____
_____
_____

Solution #2 - _____
_____
_____
_____

Solution #3 - _____
_____
_____
_____

Solution #4 - _____
_____
_____
_____

Unknown Metals

Metal A - ID prediction before rxn _____   Metal C - ID prediction before rxn _____

Metal A - ID prediction after rxn _____   Metal C - ID prediction after rxn _____

Obs._____   Obs._____

Metal B - ID prediction before rxn _____

Metal B - ID prediction before rxn _____

Obs._____

Part 2:
<u>Evaluating different solutions reacting with another</u>

1. From the fume hood, collect another well plate, familiarize yourself with the letter/number matrix of the well plate.

2. In the space provided to the right, list all of the different ionic solutions and provide a description of them. Also list the ions present in each solution and the ratio of each ion to one another - for example, Na_3PO_4 has 3 Na^+ ions and 1 PO_4^{3-} ion.

3. In the space provided below, make a matrix where each solution will be tested for solubility with one of the other solutions.

4. Once your matrix has been sketched in the space below, transfer approximately 1mL of each solution to each of the corresponding wells.

5. Allow your solutions to mix and incubate for approximately 20 minutes, gently agitate the plate to ensure the solutions mix well.

Solution	Ions	Description

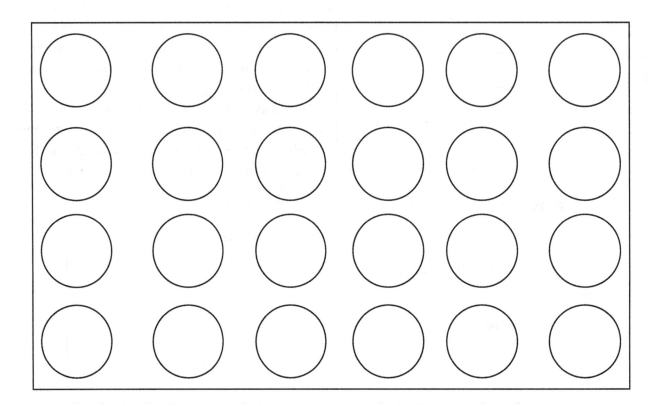

Reflections

Answer the questions below using your data - critically evaluate and analyze your results when answering these questions.

1. Did any solution combinations produce a color change?

2. Were there any solution combinations that generated a solid?

3. How many solution combinations produced a solid?

4. Based on what you know about the ions in the solutions, pick three solution combinations that generated a solid product.

5. What type of reactions are observed in part II?

6. What type of reactions are observed in part I?

7. Did your predictions of your metal identity in part I change at all?

Activity Series of Metals

Metal	Half-reaction
Lithium	$Li_{(s)} \rightarrow Li^+_{(aq)} + e^-$
Potassium	$K_{(s)} \rightarrow K^+_{(aq)} + e^-$
Barium	$Ba_{(s)} \rightarrow Ba^{2+}_{(aq)} + 2e^-$
Calcium	$Ca_{(s)} \rightarrow Ca^{2+}_{(aq)} + 2e^-$
Sodium	$Na_{(s)} \rightarrow Na^+_{(aq)} + e^-$
Magnesium	$Mg_{(s)} \rightarrow Mg^{2+}_{(aq)} + 2e^-$
Aluminum	$Al_{(s)} \rightarrow Al^{3+}_{(aq)} + 3e^-$
Manganese	$Mn_{(s)} \rightarrow Mn^{2+}_{(aq)} + 2e^-$
Zinc	$Zn_{(s)} \rightarrow Zn^{2+}_{(aq)} + 2e^-$
Chromium	$Cr_{(s)} \rightarrow Cr^{3+}_{(aq)} + 3e^-$
Iron	$Fe_{(s)} \rightarrow Fe^{2+}_{(aq)} + 2e^-$
Cobalt	$Co_{(s)} \rightarrow Co^{2+}_{(aq)} + 2e^-$
Nickel	$Ni_{(s)} \rightarrow Ni^{2+}_{(aq)} + 2e^-$
Tin	$Sn_{(s)} \rightarrow Sn^{2+}_{(aq)} + 2e^-$
Lead	$Pb_{(s)} \rightarrow Pb^{2+}_{(aq)} + 2e^-$
Hydrogen	$H_{2(g)} \rightarrow 2H^+_{(aq)} + 2e^-$
Copper	$Cu_{(s)} \rightarrow Cu^{2+}_{(aq)} + 2e^-$
Silver	$Ag_{(s)} \rightarrow Ag^+_{(aq)} + e^-$
Mercury	$Hg_{(s)} \rightarrow Hg^{2+}_{(aq)} + 2e^-$
Platinum	$Pt_{(s)} \rightarrow Pt^{2+}_{(aq)} + 2e^-$
Gold	$Au_{(s)} \rightarrow Au^{3+}_{(aq)} + 3e^-$

Look to the chemical equation - the metal on the left will be oxidized when in the presence of ions on the right IF the metal is above those ions on the activity series.

Mixing zinc metal with copper ions will result in electrons being transferred from zinc to the copper ions, forming copper metal and zinc ions.

Mixing zinc ions with copper ions will do nothing.

Mixing nickel metal with cobalt ions will do nothing. However mixing cobalt metal with nickel ions will produce cobalt ions and nickel metal.

Mixing sodium with an acid produces H_2 gas - will mixing silver with an acid also produce H_2 gas?

CHEMISTRY 1141
GENERAL CHEMISTRY LABORATORY

Experiment #8
Acid-Base Neutralization
It's all about perspective

Warnings and Hazards

The risks and hazards during this second experiment are minor however any time that you are in a lab, you must be cautious. The best way to remain safe is to understand RAMP.

Recognize hazards,
Assess risks,
Minimize risks, and
Prepare for emergencies

Using the space below, comment on each of the portions of the RAMP analysis below.

Recognize hazards (what are potential hazards that will be encountered in the procedure):

_____
_____
_____
_____

Assess risks (how can these hazards be harmful to people or your environment):

_____
_____
_____
_____

Minimize risks (what can you do to limit risk yourself and your environment):

_____
_____
_____
_____

Prepare for emergencies.
Emergency: 911
Stockroom: 512-245-3118

Background

In this experiment you will be investigating the neutralization of the strong acid, hydrochloric acid, by the strong base, sodium hydroxide. You've already looked at this reaction but this time, it'll be a little bit different.

Previously we added two aqueous solutions together - one of HCl and one of NaOH. We were trying to use a NaOH solution with a known concentration to determine the concentration of an HCl solution with an unknown concentration. In this experiment we are less concerned with the concentration of the solutions and more interested in how the temperature of the solutions changes. We are trying to figure out how much heat is produced by the reaction between NaOH and HCl.

The first reaction that we are going to be analyzing is a phase change - we will be using solid sodium hydroxide ($NaOH_{(s)}$) and adding it to deionized water. When the solid sodium hydroxide dissociates into sodium ions and hydroxide ions, is any heat released, is any heat absorbed?

The second reaction that we are going to be analyzing is the neutralization of aqueous HCl with aqueous NaOH. This reaction simply tells you what amount of heat is through the neutralization when both substances are dissociated as ions.

The third and last reaction is more or less a combination of the first and second reactions. What makes this third reaction interesting is that you will be monitoring the phase change but you will also be monitoring the neutralization. In theory the heat generated in the first reaction and the second reaction will combine to equal the heat of the third reaction.

Calculations

One of the toughest parts of this experiment is understanding how to quantify the reaction. In order to quantify this reaction we'll be using the equation shown to the right. Each of the terms for this reaction are shown below.

"q" is internal energy - the units for this are "J" or "kJ" for joules or kilojoules

$$q = mc\Delta T$$

$$\Delta H_{rxn} = q/\text{mol of limiting reactant}$$

"m" is mass - the substance which is exhibiting a temperature change/absorbing or releasing energy and the units associated with this are "g" or grams

This calculation allows us to then calculate the enthalpy (ΔH_{rxn}) change of our reaction by identifying our limiting reactant and dividing the "q" by the number of moles of our limiting reactant.

"c" is specific heat capacity - this is a constant which is unique for the substance which is exhibiting the temperature change - the units for this are joules per gram degree celsius or J/g°C

"ΔT" is change in temperature - this is calculated by taking the highest observed temperature and substracting from that the initial temperature - the units of this are °C

Example Calculation

In this experiment we will be using the specific heat capacity of deionized water for all of our calculations. Water has a specific heat capacity of 4.18J/g°C - regardless of if we are using HCl, NaOH, or deionized water we will use that value. Similarly we will use the density of water 1.0g/mL for each of our NaOH and HCl solutions.

An example calculation will look a little bit like this...

25mL of 1M HCl reacts with 25mL of 1M NaOH, the initial temperature was 23.1 °C and the final temperature was 27.8 °C, how much heat is generated in this reaction? what is the enthalpy of this reaction?

m = total volume of solutions (50mL) * density of solutions (1.0g/mL) = 50g
ΔT = final temp (27.8) - initial (23.1) = 4.7
c = specific heat capacity (4.18 J/g°C)

q= 50*4.18*4.7 = -982.3 J
sign was changed to negative because this is an exothermic reaction

Limiting reactant = both are equimolar amounts
volume of solution (25mL) * concentration (1M) = 0.025mol

982.3/0.025 = -39,292J/mol = -39.3kJ/mol

Experimental Objective

The point of this experiment is to take an experiment and break it down into components and see some of the properties of each of those pieces and add them all back together.

Bin Contents

All of your glassware will be in your bin and that bin will be collected from the stockroom after you have successfully filled out a copy of the glassware pickup document found at the back of your lab manual.

In the space below, fill in the blanks below to have a complete list of your week's glassware. In the space on the right of this page, draw each piece of glassware.

Glassware	Quantity	Capacity
_____	_____	_____
_____	_____	_____
_____	_____	_____
_____	_____	_____
_____	_____	_____
_____	_____	_____
_____	_____	_____
_____	_____	_____
_____	_____	_____
_____	_____	_____
_____	_____	_____
_____	_____	_____

Procedure
Part 1:
Heat of dissociation (solid NaOH dissolving)

1. Using the lid for your calorimeter, collect 1g of NaOH - record the mass on your data table to the right.
2. Using your 50mL graduated cylinder, collect exactly 50.0mL of dionized water and transfer it to your calorimeter (styrofoam cup).
3. Place your thermometer in the 50mL deionized water and wait for approximately 1 minute - record your initial water temperature.
4. Carefully transfer all of the solid sodium hydroxide to your calorimeter and place the lid on your calorimeter.
5. Gently swirl your calorimeter to ensure that the solid sodium hydroxide goes into solution.
6. Watch your thermometer as the temperature changes, wait approximately 3 minutes before considering the experiment complete. Wait until the temperature stops changing before you consider the reaction complete.
7. Record the final temperature in the space to the right.
8. Discard the solution in appropriate waste container, thoroughly rinse your calorimeter with deionized water and dry it with ethanol and paper towel as needed.
9. Repeat steps 1-8 two more times so that you have a total of three trials.

Trial 1
Exact mass of NaOH	_____ g
Volume of dH_2O	_____ mL
Moles of NaOH	_____
Inital temperature	_____
Final temperature	_____
Density of dH_2O	1.0 g/mL
Mass of dH_2O	_____ g
Specific heat capacity	4.18 J/g°C
Heat of reaction	_____ J

Trial 2
Exact mass of NaOH	_____ g
Volume of dH_2O	_____ mL
Moles of NaOH	_____
Inital temperature	_____
Final temperature	_____
Density of dH_2O	1.0 g/mL
Mass of dH_2O	_____ g
Specific heat capacity	4.18 J/g°C
Heat of reaction	_____ J

Trial 3
Exact mass of NaOH	_____ g
Volume of dH_2O	_____ mL
Moles of NaOH	_____
Inital temperature	_____
Final temperature	_____
Density of dH_2O	1.0 g/mL
Mass of dH_2O	_____ g
Specific heat capacity	4.18 J/g°C
Heat of reaction	_____ J

Molar conversion of NaOH

(mass NaOH)*(1mol NaOH/39.99g) = mol NaOH

Summary
Avg. Heat of rxn	_____ J
Avg. enthalpy (J/mol)	_____ J/mol

Part 2:
Heat of neutralization (NaOH and HCl solutions)

1. Thoroughly rinse your calorimeter with deionized water and dry as best as you can with ethanol and paper towels.
2. Using your 50mL graduated cylinder, collect exactly 25.0mL of 1M NaOH.
3. Using your 50mL graduated cylinder, collect exactly 25.0mL of 1M HCl.
4. Carefully transfer the 25.0mL of NaOH to your calorimeter.
5. Determine the initial temp of your NaOH solution, record in space to the left.
6. Watch your thermometer as the temperature changes, wait approximately 3 minutes before considering the experiment complete. Wait until the temperature stops changing before you consider the reaction complete.
7. Record the final temperature in the space to the right.
8. Discard the solution in appropriate waste container, thoroughly rinse your calorimeter with deionized water and dry it with paper towel.
9. Repeat steps 1-8 two more times so that you have a total of three trials.

Trial 1
- Initial temp of NaOH _____
- Final temp of solution _____
- Moles of NaOH & HCl _____
- Vol of NaOH _____
- Vol of HCl _____
- Density of NaOH and HCl ____1.0____ g/mL
- Total mass of NaOH and HCl _____
- Specific heat capacity ____4.18 J/g°C____
- Heat of reaction _____ J

Trial 2
- Initial temp of NaOH _____
- Final temp of solution _____
- Moles NaOH & HCl _____
- Vol of NaOH _____
- Vol of HCl _____
- Density of NaOH and HCl ____1.0____ g/mL
- Total mass of NaOH and HCl _____
- Specific heat capacity ____4.18 J/g°C____
- Heat of reaction _____ J

Trial 3
- Initial temp of NaOH _____
- Final temp of solution _____
- Moles of NaOH & HCl _____
- Vol of NaOH _____
- Vol of HCl _____
- Density of NaOH and HCl ____1.0____ g/mL
- Total mass of NaOH and HCl _____
- Specific heat capacity ____4.18 J/g°C____
- Heat of reaction _____ J

Summary
- Avg. Heat of rxn _____ J
- Avg. enthalpy (J/mol) _____ J/mol

Molar conversion of NaOH & HCl

vol (L) & concentration (M) = moles

Which is the limiting reactant?

$NaOH_{(aq)} + HCl_{(aq)} \rightarrow NaCl_{(aq)} + H_2O_{(l)}$

Part 3:

Heat of neutralization and dissociation (solid NaOH and HCl solutions)

1. Thoroughly rinse your calorimeter with deionized water and dry as best as you can with ethanol and paper towels.
2. Using the lid for your calorimeter, collect 1g of NaOH - record the mass on your data table to the right.
3. Using your 50mL graduated cylinder, collect exactly 25.0mL of 1M HCl.
4. Carefully transfer all of that HCl to your calorimeter.
5. Using your 50mL graduated cylinder, collect exactly 25.0mL of dH_2O and transfer that to your calorimeter.
6. Determine the initial temperature of your HCl + dH2O solution then carefully transfer the 1g of NaOH to your calorimeter.
7. Watch your thermometer as the temperature changes, wait approximately 3 minutes before considering the experiment complete. Wait until the temperature stops changing before you consider the reaction complete.
8. Record the final temperature in the space to the right.
9. Discard the solution in appropriate waste container, thoroughly rinse your calorimeter with deionized water and dry it with paper towel.
10. Repeat steps 1-9 two more times so that you have a total of three trials.

Trial 1
Inital temp of HCl + dH_2O _____
Final temp of solution _____
Mass of NaOH _____
Moles of NaOH & HCl ___ & ___
Vol of solution _____
Density of HCl ___1.0 g/mL___
Total mass of HCl + dH_2O _____
Specific heat capacity ___4.18 J/g°C___
Heat of reaction ___J___

Trial 2
Inital temp of HCl + dH_2O _____
Final temp of solution _____
Mass of NaOH _____
Moles of NaOH & HCl ___ & ___
Vol of solution _____
Density of HCl ___1.0 g/mL___
Total mass of HCl + dH_2O _____
Specific heat capacity ___4.18 J/g°C___
Heat of reaction ___J___

Trial 3
Inital temp of HCl + dH_2O _____
Final temp of solution _____
Mass of NaOH _____
Moles of NaOH & HCl ___ & ___
Vol of solution _____
Density of dH_2O + HCl ___1.0___ g/mL
Total mass of HCl + dH_2O _____
Specific heat capacity ___4.18 J/g°C___
Heat of reaction ___J___

How many moles of HCl in solution?
$C_1V_1 = C_2V_2$

Summary
Avg. Heat of rxn ___J___
Avg. enthalpy (J/mol) ___J/mol___

Analysis

1. We've looked at three related reactions to investigate Hess's law - somethign we will do again next week. The three reactions that we observed are shown below.

    1. $NaOH_{(s)} \rightarrow NaOH_{(aq)}$

    2. $NaOH_{(aq)} + HCl_{(aq)} \rightarrow NaCl_{(aq)} + H_2O_{(l)}$

    3. $NaOH_{(s)} + HCl_{(aq)} \rightarrow NaCl_{(aq)} + H_2O_{(l)}$

Notice how reaction 1 has a phase change - this is simply the sodium hydroxide dissociating into ions, in water - this is part 1 of our procedure.

Notice how reaction 2 no phase change but it has the neutralization of NaOH with HCl - this is part 2 of our procedure.

Now notice how reaction 3 has both components of 1 & 2, the phase change **AND** the neutralization - this is part 3 of our procedure.

    1. $NaOH_{(s)} \rightarrow NaOH_{(aq)}$ $\Delta H =$ _____ (J/mol)

    2. $NaOH_{(aq)} + HCl_{(aq)} \rightarrow NaCl_{(aq)} + H_2O_{(l)}$ $\Delta H =$ _____ (J/mol)

Combine the values calcuated from reactions 1 & 2 - fill in the blank below.

    3. $NaOH_{(s)} + HCl_{(aq)} \rightarrow NaCl_{(aq)} + H_2O_{(l)}$ $\Delta H =$ _____ (J/mol) (Hess's law calc)

Now - using the data from part III, calculate your experimentally observed ΔH for reaction 3.

See Part III

    4. $NaOH_{(s)} + HCl_{(aq)} \rightarrow NaCl_{(aq)} + H_2O_{(l)}$ $\Delta H =$ _____ (J/mol) (experimental)

Reflections

Answer the questions below using your data - critically evaluate and analyze your results when answering these questions.

1. Which reaction generated the least amount of heat?

2. Which reaction generated the greatest amount of heat?

3. Based on your calculations and observations - if the mass of NaOH had been increased in part 1, do you expect that the temperature change would have been greater?

4. If the concentrations of the solutions used in part 2 had been increased, do you expect that the temperature change would have been greater?

5. How did your Hess's law calculation differ from your experimentally observed ΔH_{rxn}?

6. Were all of the observed reactions exothermic or endothermic?

7. If an exothermic reaction were taking place in the palm of your hand, would your hand feel cooler or warmer?

8. What are two things that you could have changed about part 3 to produce more of a temperature change?

9. Is it possible to reverse an exothermic reaction - if so, how?

CHEMISTRY 1141
GENERAL CHEMISTRY LABORATORY

Experiment #9
Obey Hess's Law
It Can't be Broken

Warnings and Hazards

The risks and hazards during this second experiment are minor however any time that you are in a lab, you must be cautious. The best way to remain safe is to understand RAMP.

    Recognize hazards,
    Assess risks,
    Minimize risks, and
    Prepare for emergencies

Using the space below, comment on each of the portions of the RAMP analysis below.

Recognize hazards (what are potential hazards that will be encountered in the procedure):

_____
_____
_____
_____

Assess risks (how can these hazards be harmful to people or your environment):

_____
_____
_____
_____

Minimize risks (what can you do to limit risk yourself and your environment):

_____
_____
_____
_____

Prepare for emergencies.
    Emergency: 911
    Stockroom: 512-245-3118

Background

This experiment will be investigating a topic similar to last week's however we'll be taking a deeper dive into Hess's Law. Last week we were primarily looking at a reaction as individual pieces. We looked at the different angles of a reaction and this week we're going to be looking at a new and different reaction to analyze Hess's law.

Hess's Law was originally published in 1840 and since then, it remains undefeated! Essentially, Hess's law theorizes that the enthalpy change of a reaction is the same if a reaction takes place in a series of steps or through one step. An example of Hess's Law is shown below.

$$CH_{4(g)} + 2O_{2(g)} \rightarrow CO_{2(g)} + \cancel{2H_2O_{(g)}} \quad \Delta H_{rxn} = -802 kJ$$

$$\cancel{2H_2O_{(g)}} \rightarrow 2H_2O_{(l)} \quad \Delta H_{rxn} = -88 kJ$$

$$\overline{CH_{4(g)} + 2O_{2(g)} \rightarrow CO_{2(g)} + 2H_2O_{(l)} \quad \Delta H_{rxn} = -890 kJ}$$

The example above shows the condensation of gaseous H_2O converting to liquid H_2O and how that phase change impacts the overall enthalpy of the reaction.

We will be investigating the reaction of solid magnesium's reaction with O_2 to form solid magnesium oxide. We'll be investigating this reaction through magnesium's reaction with HCl as well as magnesium oxide's reaction with HCl. We'll be able to evaluate our work with those two and another reaction to determine the ΔH_{rxn} of MgO.

Experimental Objective

The point of this experiment is to dissect a chemical reaction and look at how it can be completed in multiple different steps. Essentially, we want to use this experiment and last week's experiment to investigate Hess's Law but also to validate/invalidate it!

Bin Contents

All of your glassware will be in your bin and that bin will be collected from the stockroom after you have successfully filled out a copy of the glassware pickup document found at the back of your lab manual.

In the space below, fill in the blanks below to have a complete list of your week's glassware. In the space on the right of this page, draw each piece of glassware.

Glassware	Quantity	Capacity
_____	_____	
_____	_____	
_____	_____	
_____	_____	
_____	_____	
_____	_____	
_____	_____	
_____	_____	
_____	_____	
_____	_____	
_____	_____	
_____	_____	

Procedure
Part 1:
Reacting HCl with Mg

1. Balance the chemical reaction at the bottom of the page.
2. Using your 50mL graduated cylinder, collect 25.0mL of HCl and add it to your calorimeter.
3. Using your 50mL graduated cylinder, collect 25.0mL of dH_2O and add it to your calorimeter containing the HCl.
4. Place your thermometer in the 50mL HCl solution and wait for approximately 1 minute - record the initial water temperature.
5. Using your calorimeter lid, collect approximately 0.3g of Mg, record the precise mass in the space to the left.
6. Carefully transfer the solid Mg to your calorimeter.
7. Gently swirl your calorimeter to ensure that the solid completely reacts with the HCl solution.
8. Watch your thermometer as the temperature changes, wait approximately 3 minutes before considering the experiment complete. Wait until the temperature stops changing before you consider the reaction complete.
9. Record the final temperature in the space to the right.
10. Discard the solution in appropriate waste container, thoroughly rinse your calorimeter with deionized water and dry it with ethanol and paper towel as needed.
11. Repeat steps 1-8 two more times so that you have a total of three trials.

__ $HCl_{(aq)}$ + __ $Mg_{(s)}$ -> __ $MgCl_{2(aq)}$ + __ $H_{2(g)}$

Trial 1
- Mass of Mg   _____ g
- Moles of Mg   _____
- Volume of HCl   _____
- Moles of HCl   _____
- Limiting reagent   __HCl or Mg__
- Inital temperature   _____
- Final temperature   _____
- Density of HCl + dH_2O __1.0 g/mL__
- Mass of HCl + dH_2O _____ g
- Specific heat capacity   __4.18 J/g°C__
- Heat of reaction   _____ J

Trial 1
- Mass of Mg   _____ g
- Moles of Mg   _____
- Volume of HCl   _____
- Moles of HCl   _____
- Limiting reagent   __HCl or Mg__
- Inital temperature   _____
- Final temperature   _____
- Density of HCl + dH_2O __1.0 g/mL__
- Mass of HCl + dH_2O _____ g
- Specific heat capacity   __4.18 J/g°C__
- Heat of reaction   _____ J

Trial 1
- Mass of Mg   _____ g
- Moles of Mg   _____
- Volume of HCl   _____
- Moles of HCl   _____
- Limiting reagent   __HCl or Mg__
- Inital temperature   _____
- Final temperature   _____
- Density of HCl + dH_2O __1.0 g/mL__
- Mass of HCl + dH_2O _____ g
- Specific heat capacity   __4.18 J/g°C__
- Heat of reaction   _____ J

Summary
- Avg. Heat of rxn   _____ J
- Avg. enthalpy (J/mol)   _____
- Avg. enthalpy (kJ/mol)   _____

Part 2:
Reacting HCl with MgO

1. Balance the chemical reaction at the bottom of the page.
2. Using your 50mL graduated cylinder, collect 25.0mL of HCl and add it to your calorimeter.
3. Using your 50mL graduated cylinder, collect 25.0mL of dH_2O and add it to your calorimeter containing the HCl.
4. Place your thermometer in the 50mL HCl solution and wait for approximately 1 minute - record the initial water temperature.
5. Using your calorimeter lid, collect approximately 0.3g of Mg, record the precise mass in the space to the left.
6. Carefully transfer the solid MgO to your calorimeter.
7. Gently swirl your calorimeter to ensure that the solid completely reacts with the HCl solution.
8. Watch your thermometer as the temperature changes, wait approximately 3 minutes before considering the experiment complete. Wait until the temperature stops changing before you consider the reaction complete.
9. Record the final temperature in the space to the right.
10. Discard the solution in appropriate waste container, thoroughly rinse your calorimeter with deionized water and dry it with ethanol and paper towel as needed.
11. Repeat steps 1-8 two more times so that you have a total of three trials.

Trial 1
Mass of MgO _____ g
Moles of MgO _____
Volume of HCl _____
Moles of HCl _____
Limiting reagent ___HCl or MgO___
Inital temperature _____
Final temperature _____
Density of HCl + dH_2O ___1.0 g/mL___
Mass of HCl + dH_2O _____ g
Specific heat capacity ___4.18 J/g°C___
Heat of reaction _____ J

Trial 1
Mass of MgO _____ g
Moles of MgO _____
Volume of HCl _____
Moles of HCl _____
Limiting reagent ___HCl or MgO___
Inital temperature _____
Final temperature _____
Density of HCl + dH_2O ___1.0 g/mL___
Mass of HCl + dH_2O _____ g
Specific heat capacity ___4.18 J/g°C___
Heat of reaction _____ J

Trial 1
Mass of MgO _____ g
Moles of MgO _____
Volume of HCl _____
Moles of HCl _____
Limiting reagent ___HCl or MgO___
Inital temperature _____
Final temperature _____
Density of HCl + dH_2O ___1.0 g/mL___
Mass of HCl + dH_2O _____ g
Specific heat capacity ___4.18 J/g°C___
Heat of reaction _____ J

Summary
Avg. Heat of rxn _____ J
Avg. enthalpy (J/mol) _____
Avg. enthalpy (kJ/mol) _____

Hess's Law is all about utilizing a series of reactions and their ΔH to determine the ΔH of an unknown reaction.

1. $Mg_{(s)} + 2HCl_{(aq)} \rightarrow MgCl_{2(aq)} + H_{2(g)}$ ΔH = _____ kJ/mol

2. $MgO_{(s)} + 2HCl_{(aq)} \rightarrow MgCl_{2(aq)} + H_2O_{(l)}$ ΔH = _____ kJ/mol

3. $H_{2(g)} + \frac{1}{2}O_{2(g)} \rightarrow H_2O_{(l)}$ ΔH = -285.8 kJ/mol

4. $Mg_{(s)} + \frac{1}{2}O_{2(g)} \rightarrow MgO_{(s)}$ ΔH = _____ kJ/mol

In the space below, rewrite reactions 1-3 (if needed) so that they will align with reaction 4 - be sure to write the ΔH values and adjust the signs accordingly.

ΔH = _____ kJ/mol

ΔH = _____ kJ/mol

ΔH = _____ kJ/mol

Reflections

Answer the questions below using your data - critically evaluate and analyze your results when answering these questions.

1. Which reaction generated the least amount of heat?

2. Which reaction generated the greatest amount of heat?

3. Is the production of magnesium oxide, endothermic or exothermic?

4. How would your results change if you were to double the concentration of the HCl solution?

5. What type of reaction is taking place between MgO and HCl?

6. What type of reaction is taking place between Mg and HCl?

7. In the reactions of MgO and Mg with HCl - is any oxidation or reduction taking place?

8. Is it possible to produce Mg from MgCl₂ and H₂ - if so, how would that impact your ΔH? Would this be an endothermic or exothermic reaction?

CHEMISTRY 1141
GENERAL CHEMISTRY LABORATORY

Experiment #10
This Experiment?
It's a Gas!

Warnings and Hazards

The risks and hazards during this second experiment are minor however any time that you are in a lab, you must be cautious. The best way to remain safe is to understand RAMP.

Recognize hazards,
Assess risks,
Minimize risks, and
Prepare for emergencies

Using the space below, comment on each of the portions of the RAMP analysis below.

Recognize hazards (what are potential hazards that will be encountered in the procedure):

_____
_____
_____
_____
_____

Assess risks (how can these hazards be harmful to people or your environment):

_____
_____
_____
_____
_____

Minimize risks (what can you do to limit risk yourself and your environment):

_____
_____
_____
_____
_____

Prepare for emergencies.
    Emergency: 911
    Stockroom: 512-245-3118

Background

This experiment is to help us gain a greater understanding of the ideal gas law. This experiment will help you work with the ideal gas equation as well!

In this experiment you will be reacting HCl with a compound which will generate a carbonate ion which will break down to produce carbon dioxide gas. The carbon dioxide gas will be captured and the volume of that gas will be measured. Ultimately you will be calculating "R" which is sometimes known as the molar gas constant, the gas constant, or the ideal gas constant. This constant is an example of a proportionality constant which ties in observations of Avogadro's law, Boyle's law, Charles's law, and Gay-Lussac's law. The ideal gas equation and the different constants are shown below.

$$PV = nRT$$

P - pressure, atm
V - volume, L
n - moles of gas, mol
T - temperature, K

R is the value that we will be solving for, other values for this equation will measured in the lab. Our aim and goal with this experiment is to calculate R based on our experiment and compare that to the R-value that you'll find in virtually any chemistry textbook - 0.0821 L atm/mol K.

Experimental Objective

The point of this experiment is to learn observe an experiment where gas can be generated as a product and that product can be measured. This experiment also serves as a way to experimentally the R value from the ideal gas equation.

Bin Contents

All of your glassware will be in your bin and that bin will be collected from the stockroom after you have successfully filled out a copy of the glassware pickup document found at the back of your lab manual.

In the space below, fill in the blanks below to have a complete list of your week's glassware. In the space on the right of this page, draw each piece of glassware.

Glassware	Quantity	Capacity

Procedure
Part 1:
Reacting HCl with NaHCO₃

1. Balance the chemical reaction at the bottom of the page.
2. Fill one of the pitchers in your lab to ~1.75L.
3. Fill your 100mL grad cylinder to the top.
4. Invert your grad cylinder into the pitcher.
5. Read the grad cylinder to learn and note the volume of gas in the cylinder.
6. Carefully thread tubing from the stopper into the open end of the grad cylinder.
7. Collect 1.00g of sodium bicarbonate and transfer it into the round-bottom flask.
8. Attach round bottom flask to the 3-way claisen adapter - add clamp to secure it.
9. Secure the stopper with the tubing into 3-way adapter.
10. Carefully collect 15.0mL of HCl in your grad cylinder.
11. Carefully transfer all HCl into round-bottom flask via opening of 3-way adapter and quickly cover opening with black stopper.
12. Watch as gas begins to accumulate in inverted graduated cylinder.
13. When bubbles have stopped flowing into graduated cylinder, record the final volume of the gas in the graduated cylinder.
14. Fill in the data table to the left.
15. Carefully rinse and clean all of the glassware and repeat steps 3-14 two more times.

Trial 1
Barometric Pressure _____ atm
Temperature _____ K
Volume of HCl _____
Concentration of HCl _____ g
Moles of HCl _____
Mass of NaHCO₃ _____
Moles of NaHCO₃ _____
Initial volume _____
Final volume _____
Total volume of CO_2 _____
Limiting Reagent _____
Moles of CO_2 generated _____
Experimental R value _____

Trial 2
Barometric Pressure _____ atm
Temperature _____ K
Volume of HCl _____
Concentration of HCl _____ g
Moles of HCl _____
Mass of NaHCO₃ _____
Moles of NaHCO₃ _____
Initial volume _____
Final volume _____
Total volume of CO_2 _____
Limiting Reagent _____
Moles of CO_2 generated _____
Experimental R value _____

Trial 3
Barometric Pressure _____ atm
Temperature _____ K
Volume of HCl _____
Concentration of HCl _____ g
Moles of HCl _____
Mass of NaHCO₃ _____
Moles of NaHCO₃ _____
Initial volume _____
Final volume _____
Total volume of CO_2 _____
Limiting Reagent _____
Moles of CO_2 generated _____
Experimental R value _____

___$HCl_{(aq)}$ + ___$NaHCO_{3(s)}$ -> ___$NaCl_{(aq)}$ + ___$CO_{2(g)}$ + ___$H_2O_{(l)}$

Summary
Avg. Experimental R value _____

Reflections

Answer the questions below using your data - critically evaluate and analyze your results when answering these questions.

1. What is the lowest possible temperature Kelvin?

2. Does the identity of the gas matter when doing a calculation with the ideal gas law?

3. What source(s) of error are present when doing this experiment?

4. What was the limiting reagent in this experiment?

5. Would your results change if you used 3M HCl as opposed to 1M HCl?

6. What ions does $NaHCO_3$ breakdown into?

7. What other reactions have we carried out this semester that generate gas?

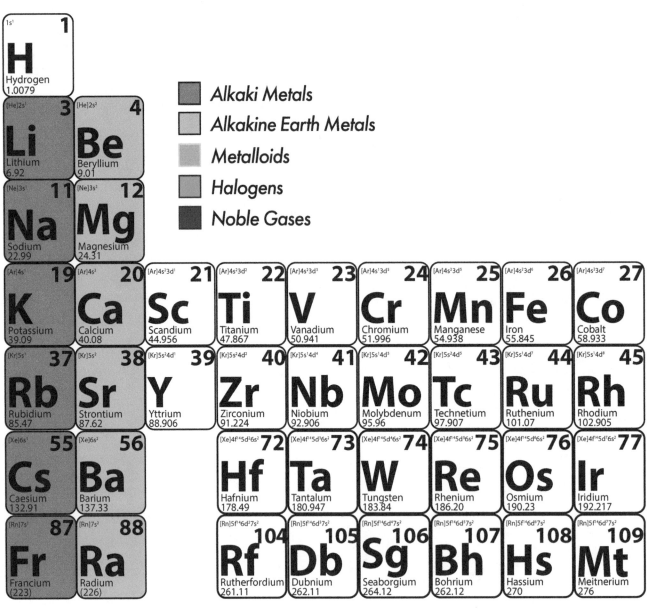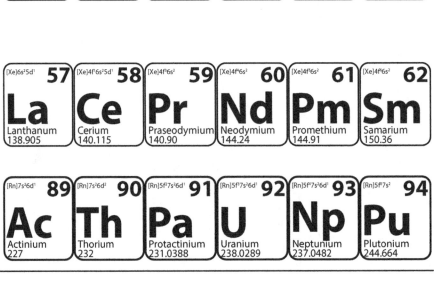

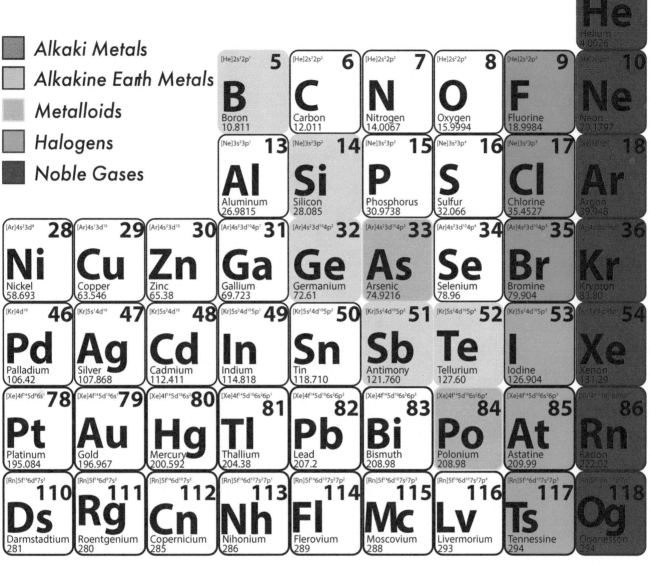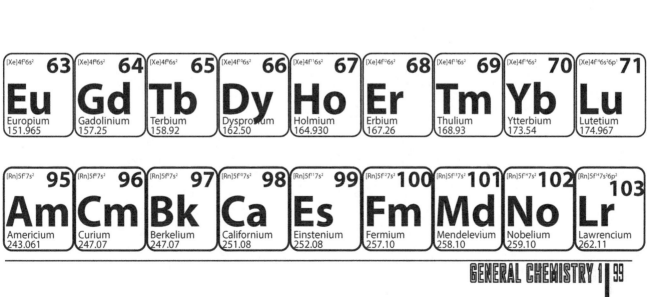

Bobcat Chemistry

Chemistry & Biochemistry Stockroom Supply Form

THIS IS YOUR RECEIPT! KEEP FOR YOUR RECORDS!

ALL PARTIES MUST FILL THE FORM OUT COMPLETELY TO RECEIVE SUPPLIES.

A Photo ID is REQUIRED to receive any supplies.

Lost/broken items will result in a feee assessed to your student account in the full amount of items checked out.

Failure to return items will result in a $50 fee assessed to your student account.

DATE: __ __ / __ __ / __ __   COURSE NUMBER: __ __ __ __   SECTION NUMBER: __ __ __

_____   _____   A _ _ _ _ _ _ _
Last Name   First Name   Student ID

TA Name: _____   TA Initials: _____

FOR COURSES THAT WORK IN GROUPS ONLY

_____   _____   A _ _ _ _ _ _ _
2ND Student Last Name   2ND Student First Name   2ND Student ID

_____   _____   A _ _ _ _ _ _ _
3RD Student Last Name   3RD Student First Name   3RD Student ID

STUDENTS WHO WORK IN GROUPS WILL HAVE ANY AND ALL SUPPLY CHARGES DISTRIBUTED AMONG ALL PARTIES

SUPPLY ITEMS ISSUED TO STUDENT (STOCKROOM INITIALS)

ITEM NAME	CHECKED OUT	CHECKED IN
1) _____	_____	_____
2) _____	_____	_____
3) _____	_____	_____
4) _____	_____	_____
5) _____	_____	_____

GENERAL CHEMISTRY 1 | 101

Bobcat Chemistry

Chemistry & Biochemistry Stockroom Supply Form

THIS IS YOUR RECEIPT! KEEP FOR YOUR RECORDS!

ALL PARTIES MUST FILL THE FORM OUT COMPLETELY TO RECEIVE SUPPLIES.

A Photo ID is REQUIRED to receive any supplies.

Lost/broken items will result in a feee assessed to your student account in the full amount of items checked out.

Failure to return items will result in a $50 fee assessed to your student account.

DATE: __ __ / __ __ / __ __   COURSE NUMBER: __ __ __ __   SECTION NUMBER: __ __ __

_____   _____   A __ __ __ __ __ __ __
Last Name   First Name   Student ID

TA Name: _____   TA Initials: _____

FOR COURSES THAT WORK IN GROUPS ONLY

_____   _____   A __ __ __ __ __ __ __
2ND Student Last Name   2ND Student First Name   2ND Student ID

_____   _____   A __ __ __ __ __ __ __
3RD Student Last Name   3RD Student First Name   3RD Student ID

STUDENTS WHO WORK IN GROUPS WILL HAVE ANY AND ALL SUPPLY CHARGES DISTRIBUTED AMONG ALL PARTIES

SUPPLY ITEMS ISSUED TO STUDENT (STOCKROOM INITIALS)

ITEM NAME	CHECKED OUT	CHECKED IN
1) _____	_____	_____
2) _____	_____	_____
3) _____	_____	_____
4) _____	_____	_____
5) _____	_____	_____

GENERAL CHEMISTRY 1 | 103

Bobcat Chemistry

Chemistry & Biochemistry Stockroom Supply Form

THIS IS YOUR RECEIPT! KEEP FOR YOUR RECORDS!

ALL PARTIES MUST FILL THE FORM OUT COMPLETELY TO RECEIVE SUPPLIES.

A Photo ID is REQUIRED to receive any supplies.

Lost/broken items will result in a feee assessed to your student account in the full amount of items checked out.

Failure to return items will result in a $50 fee assessed to your student account.

DATE: __ __ / __ __ / __ __   COURSE NUMBER: __ __ __ __   SECTION NUMBER: __ __ __

_____   _____   A __ __ __ __ __ __ __
Last Name   First Name   Student ID

TA Name: _____   TA Initials: _____

FOR COURSES THAT WORK IN GROUPS ONLY

_____   _____   A __ __ __ __ __ __ __
2ND Student Last Name   2ND Student First Name   2ND Student ID

_____   _____   A __ __ __ __ __ __ __
3RD Student Last Name   3RD Student First Name   3RD Student ID

STUDENTS WHO WORK IN GROUPS WILL HAVE ANY AND ALL SUPPLY CHARGES DISTRIBUTED AMONG ALL PARTIES

SUPPLY ITEMS ISSUED TO STUDENT (STOCKROOM INITIALS)

ITEM NAME	CHECKED OUT	CHECKED IN
1) _____	_____	_____
2) _____	_____	_____
3) _____	_____	_____
4) _____	_____	_____
5) _____	_____	_____

GENERAL CHEMISTRY 1 | 105

Bobcat Chemistry

Chemistry & Biochemistry Stockroom Supply Form

THIS IS YOUR RECEIPT! KEEP FOR YOUR RECORDS!

ALL PARTIES MUST FILL THE FORM OUT COMPLETELY TO RECEIVE SUPPLIES.

A Photo ID is REQUIRED to receive any supplies.

Lost/broken items will result in a feee assessed to your student account in the full amount of items checked out.

Failure to return items will result in a $50 fee assessed to your student account.

DATE: __ __ / __ __ / __ __   COURSE NUMBER: __ __ __ __   SECTION NUMBER: __ __ __

_____   _____   A _ _ _ _ _ _ _
Last Name   First Name   Student ID

TA Name: _____   TA Initials: _____

FOR COURSES THAT WORK IN GROUPS ONLY

_____   _____   A _ _ _ _ _ _ _
2ND Student Last Name   2ND Student First Name   2ND Student ID

_____   _____   A _ _ _ _ _ _ _
3RD Student Last Name   3RD Student First Name   3RD Student ID

STUDENTS WHO WORK IN GROUPS WILL HAVE ANY AND ALL SUPPLY CHARGES DISTRIBUTED AMONG ALL PARTIES

SUPPLY ITEMS ISSUED TO STUDENT (STOCKROOM INITIALS)

ITEM NAME	CHECKED OUT	CHECKED IN
1) _____	_____	_____
2) _____	_____	_____
3) _____	_____	_____
4) _____	_____	_____
5) _____	_____	_____

GENERAL CHEMISTRY 1 | 107

Bobcat Chemistry

Chemistry & Biochemistry Stockroom Supply Form

THIS IS YOUR RECEIPT! KEEP FOR YOUR RECORDS!

ALL PARTIES MUST FILL THE FORM OUT COMPLETELY TO RECEIVE SUPPLIES.

A Photo ID is REQUIRED to receive any supplies.

Lost/broken items will result in a feee assessed to your student account in the full amount of items checked out.

Failure to return items will result in a $50 fee assessed to your student account.

DATE: __ __ / __ __ / __ __   COURSE NUMBER: __ __ __ __   SECTION NUMBER: __ __ __

_____   _____   A __ __ __ __ __ __ __
Last Name   First Name   Student ID

TA Name: _____   TA Initials: _____

FOR COURSES THAT WORK IN GROUPS ONLY

_____   _____   A __ __ __ __ __ __ __
2ND Student Last Name   2ND Student First Name   2ND Student ID

_____   _____   A __ __ __ __ __ __ __
3RD Student Last Name   3RD Student First Name   3RD Student ID

STUDENTS WHO WORK IN GROUPS WILL HAVE ANY AND ALL SUPPLY CHARGES DISTRIBUTED AMONG ALL PARTIES

SUPPLY ITEMS ISSUED TO STUDENT (STOCKROOM INITIALS)

ITEM NAME	CHECKED OUT	CHECKED IN
1) _____	_____	_____
2) _____	_____	_____
3) _____	_____	_____
4) _____	_____	_____
5) _____	_____	_____

GENERAL CHEMISTRY 1 | 109